Vol. 1
BASIC SCIENTIFIC SUBROUTINES

Biology Dept.
Northeastern Ill. Univ
3-81

Biology Dept.
Northeastern Ill. Univ.
3-81

Vol.1
BASIC SCIENTIFIC SUBROUTINES

by
F. R. Ruckdeschel

BYTE/McGRAW-HILL
70 MAIN ST
PETERBOROUGH, NH 03458

BASIC SCIENTIFIC SUBROUTINES Vol. 1

Copyright © 1981 by BYTE Publications Inc. All rights reserved. Printed in the United States of America. No part of this publication may be reproduced, stored in a retrieval system, or transmitted, in any form or by any means, electronic, mechanical, photocopying, recording, or otherwise, without the prior written permission of the publisher.

The author of the programs provided with this book has carefully reviewed them to ensure their performance in accordance with the specifications described in the book. Neither the author nor BYTE Publications Inc, however, make any warranties concerning the programs and assume no responsibility or liability of any kind for errors in the programs, or for the consequences of any such errors. The programs are the sole property of the author and have been registered with the United States Copyright Office.

Library of Congress Cataloging in Publication Data

Ruckdeschel, F R

    Basic scientific subroutines.

    Bibliography: v.1, p.
    1. Mathematics — Computer programs. 2. Basic (Computer program language) I. Title.
QA76.95.R82   502.8'5425   80-19582
ISBN 0-07-054201-5 (v.1)

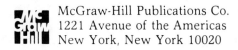
McGraw-Hill Publications Co.
1221 Avenue of the Americas
New York, New York 10020

ACKNOWLEDGEMENTS

The author is grateful to Carl Helmers and his colleagues for offering the opportunity to present this information to other computer users. The author also wishes to acknowledge the helpful criticism and proofreading of the manuscript by A. Walsh, J. Krinsky, and O. Hauser.

Table of Contents

Preface

Chapter I   Introduction   **1**

1. Scope 1
2. Contents 2
3. Structure 2
4. Requirements 3
5. Execution Speed 4

Chapter II   Plotting Subroutines   **5**

1. One-Dimensional Data Plot 6
2. Function Plotting 13
3. Two-Dimensional Data Plot 16

Chapter III   Complex Variables   **24**

1. The Complex Plane 24
2. Complex Variable Operations 31
3. Powers and Roots of $Z=X+iY$ 38
4. Spherical Coordinate Conversion 47

Chapter IV   Vector and Matrix Operations   **53**

1. Vector Operations 53
2. Matrix Sums and Products 63

3. Other Matrix Operations 79
4. Matrix Coordinate Changes 86
5. Determinants 87
6. Matrix Inversion 94
7. Solving Linear Sets of Equations 101
8. Characteristic Polynomials and Eigenvalues 105
9. Eigenvalues by the Power Method 111
10. Matrix Exponentiation and Differential Equations 117

Chapter V   Random Number Generators   125

1. The Uniform Distribution, RND 127
2. The Linear Distribution 131
3. The Normal Distribution 134
4. The Poisson Distribution 137
5. The Binomial Distribution 141
6. The Exponential Distribution 144
7. The Fermi Distribution 146
8. The Cauchy Distribution 148
9. The Gamma Distribution 150
10. The Beta Distribution 152
11. The Weibull Distribution 155

Chapter VI   Basic Series Approximations   159

1. Taylor Series Expansions 159
2. Approximate Series Expansions 162
3. Variations on the Optimal Series Theme 165
4. Least-Squares Regression 167
5. Extensions 191

Appendix I   Subroutine Cross Index   193
A. Software Index (by number)
B. Function Index

Appendix II   Subroutine Listings   219
A. Full Listings of Demonstration and Subroutine Programs
B. Compacted Listings

Appendix III Conversion to Other BASICs and Microsoft BASIC
Program Listings   265

References   311

Index   315

PREFACE

BASIC is rapidly becoming one of the major computer languages in use. It was developed by Professor J. G. Kemeny and others at Dartmouth College for the purpose of introducing computer programming at a more universal skill level than required by other languages, such as FORTRAN. BASIC has met with great success because of its simple syntax and conversational nature. Various dialects may be found in large time-sharing computer systems, in minicomputers, and most recently, in microcomputers. It is likely that such use will continue to expand in the future, particularly in the microcomputer realm.

Dedicated microcomputers containing BASIC software offer the possibility of personal computing on a scale scarcely envisioned only a few years ago. They also offer a convenient and cost-effective alternative to large system time-sharing. In heavy use situations, the dedicated microcomputer can offer faster response as well as lower computing cost. The two key disadvantages are that BASIC is generally implemented as an interpreter and thus executes slowly, and that the BASIC user generally has available only a limited set of elementary functions compared to the extensive subroutine libraries of large time-sharing systems. A language such as FORTRAN is at an advantage in this respect.

FORTRAN source code is assembled as a whole into a run-time package consisting of optimized object code, the language used by the computer. This object code is relatively fast in execution. BASIC is usually assembled statement by statement as each is encountered. Thus FORTRAN executes much more rapidly. However, in mathematics-intensive operations, BASIC and FORTRAN may differ in speed by only a factor of two to ten, since both access similar time-consuming floating-point subroutines.

In existence for some time, FORTRAN has been extensively used by the scientific community. Accordingly, large subroutine libraries have been developed, such as SPSS (Statistical Package for the Social Sciences; reference 24). In many cases it would not be advantageous for a user to give up the inherent speed and extensive software library of FORTRAN on a large computer for BASIC on a microcomputer.

However, these two particular disadvantages of BASIC will be lessened in the future. A BASIC compiler will eventually become an option on many microcomputers. The user will then be able to take advantage of the conversational features of a BASIC interpreter for program development and also exercise the speed advantage of a BASIC compiler. Also, scientific subroutines are being written for BASIC applications at an increasing rate. This particular book represents one such contribution.

Several considerations went into preparing this book. The first was scope. As a physicist and engineer for a major US corporation, the author has been presented with a spectrum of technical assignments, each requiring different skills and techniques. Based on that experience and an examination of the analysis methods being presented in current college courses, the subjects appearing in the table of contents were chosen. This list is certainly not complete. For example, statistical techniques and their associated subroutines are hardly addressed. Future volumes will extend the subroutine library to include statistical routines and more.

The form of presentation was another consideration. As the subroutine library is expected to be used by a wide range of individuals, including high school and college students, as well as practicing professionals, simple program listings were considered to be insufficient. The format chosen begins with a discussion of possible subroutine uses, followed by a brief mathematical description, and concludes with the subroutine listing along with examples. References are supplied so that the advanced programmer can learn more about the procedures used and possibly expand upon them. In some cases, suggestions are given with regard to further applications which are beyond the scope of the present book, but may be of use to some readers.

Another concern in preparing this book was which BASIC dialect to use. There are several dialects, including those supported by Microsoft, Inc. and North Star. For this book, a subset (with some exceptions) of both Microsoft and North Star BASIC was chosen in order to be compatible with as many systems as possible. Several manufacturers of microcomputers supply BASIC interpreters under their own trade names, even though in actuality the source is Microsoft. For example, Apple, Commodore, Ohio Scientific, and Radio Shack are in this class. Thus, using a subset of these major dialects permits ready transference of the subroutines presented to most computers with little revision. The disadvantage of not using all the capability of Microsoft or North Star BASIC is reflected in program size and efficiency. For example, multiple statement lines are not used, since the statement delimiter characters may be incompatible with other BASIC interpreters.

To assist the user, the North Star-based subroutines presented in the main text are also given in concatenated form in the appendix. Compacted North Star subroutines are also presented to improve memory utilization. The software appearing in this book is offered elsewhere in several machine-readable forms, including cassette, paper tape, and disk. Finally, Appendix III provides information on converting the routines given to other forms of BASIC.

Considerable attention has been paid to presenting the information contained in this book in a form that is readily usable by the many individuals who have access to microcomputers containing a BASIC interpreter. Hopefully, the application of such systems will be enhanced by this material.

F. R. Ruckdeschel

Vol.1
BASIC SCIENTIFIC SUBROUTINES

CHAPTER I:

INTRODUCTION

I.1 Scope

BASIC Scientific Subroutines is a coordinated collection of programs written in BASIC which provides many of the library functions common to large time-sharing computer systems. Most of the subroutines presented are based on standard methods found in texts on applied mathematics, numerical methods, and statistics. These analysis techniques have been converted into computer programs using a language subset of North Star and/or Microsoft BASIC.

Each subroutine is documented with a description of the class of problems addressed by that program, an outline of the mathematical principles, a program listing, and at least one example. In some cases alterations that extend the use of the routine are discussed. In all cases every effort has been made to clearly present the information, ensuring minimal difficulty in implementation.

Although this book is not intended to be a complete software library, it does provide a spectrum of programs of interest to many scientific programmers. The material is written with several audiences in mind, ranging from the engineering student who is just learning how to program to the professional who wishes to more fully understand the sources of the techniques used.

This text is not meant to provide a tutorial on programming in BASIC. However, the programs listed are extensively documented internally through the liberal use of REMark statements. This permits the user to follow the logic of the program and perhaps make alterations to suit a particular need. However, this internal documentation has a negative effect on the amount of memory required and on execution speed. To overcome this, all programs presented in the main text are reproduced in Appendix II in compacted form.

This book is the first in a series on scientific subroutines. The present text, Volume I, deals with plotting routines, matrix mathematics, and approximation techniques. Volume II considers approximation, regression, interpolation, integration, root-seeking, and optimization. These are subjects common to many of the sciences.

I.2 Contents

In preparing this book, choices had to be made regarding content. The content was ultimately determined by an estimate of the estimated use of the routines and their relevance to common scientific computing needs.

Plotting routines are always needed, and Chapter II addresses graphical displays in several ways. Each routine recognizes the fact that most programmers have only limited access to high-resolution XY plotters. Thus, the programs given are designed to utilize the limited capability of the character-oriented terminals commonly used for direct programming and printout.

Chapter III focuses on complex variables. In one case, the rules of complex algebra are applied to converting between rectangular and polar (cylindrical, polar, and spherical) coordinates. In another case, the powers of $Z = X + iY$ are considered. The subroutine presented for that problem has further utility in the polynomial expansion of analytic functions.

Chapter IV contains several subroutines pertaining to matrix operations. It provides functions generally not found in BASIC interpreters for microcomputers, such as matrix inversion. Also presented are equation-solving techniques which employ matrix algebra methods.

Chapter V considers one of the most important functions used in computer simulations, the random number generator. Subroutines are supplied for generating both uniform and normally distributed random number sequences. Many of the common BASIC interpreters contain uniformly distributed random number functions, but few, if any, offer any others. The Normal distribution in particular is useful in many applications, such as Monte Carlo analysis and signal/noise evaluations.

Series approximations are dealt with in Chapter VI in a preliminary fashion. Several important functions are approximated using series which slightly differ from the Taylor expansions often used by computer programmers. The techniques presented tend to be high in accuracy and fast in execution. Also discussed is the generation of other series approximations to suit particular needs.

I.3 Structure

Each subroutine presentation is laid out in a particular pattern. First there is a discussion of the scientific use of the subroutine, followed by an outline of the mathematical techniques which form the basis of the computer code. In many cases, the mathematical outline will be of interest to the reader by itself, since it represents a digestion of a more complete discussion to be found in the cited references. Next comes a description of the input and output characteristics of the subroutine, including the memory requirements and approximate execution speed. Also given are the functions and variables used within the subroutine code. A listing of the subroutine is provided, followed by examples. In some cases, extensions of the subroutine to other or more complex problems are considered along with cautions regarding blanket application of the program. The latter discussion is important because situations can arise in which even the most well-written program fails to operate as expected.

The above structure was chosen after reviewing a few of the current books on software. Some books dwell on the mathematics of the technique with little discussion of probable use. Other books are primarily program listings with little background. Interestingly, some of the most well-rounded texts are those written for programmable calculators, in particular the Hewlett-Packard series (references 4, 5, 12, 13, and 26). In some cases the classical algorithms presented in programmable calculator handbooks are nearly ideal for computer implementation, particularly those involving iteration techniques.

I.4 Requirements

The specific hardware and software requirements are listed in each subsection for the program being considered. Most of the subroutines presented require less than 2 K bytes of memory above the size of the operating system. In several of the programs involving dimensioned arrays of data, the memory demand is dictated by the size of the data set and can therefore be quite large.

An attempt was made to keep the terminal line width requirements below thirty-two characters, though this was not possible in some of the tabular displays. In almost all cases, a sixty-four-character (uppercase) display is sufficient. All demonstration program input data is assumed to be entered from the keyboard. All execution time comparisons are based on an IMSAI 8080 running at 2 MHz with no wait states.

The full set of statements and operations available with Microsoft and North Star BASIC was not fully utilized in order to permit wider use of the subroutines. In particular, multiple statement lines are not used, and the input/output formats are kept very simple. Few functions are assumed, and even those can be replaced by subroutines found elsewhere in the text. For example, the trigonometric functions are used in some routines, but series expansions for them are also provided.

The programs are presented as subroutines; that is, they are terminated with a RETURN statement. Since it might be desirable to assemble the software into a subroutine library package, the routines are sequentially numbered, starting at statement 40000. This places them high enough in line number to be out of the way of the main calling program.

When using the subroutines, pay attention to overlaps between the variable list used by the main program and that employed by the subroutine. The following conventions have been chosen:

- **FOR/NEXT loops:** Six general indices have been reserved: I<n>, J<n>, K<n>, L<n>, M<n>, and N<n>, where <n> is optionally a digit in the se-

cond location. Included in this convention are the dimensioned extensions. For example, I4(K) might be found in a subroutine.
- **Running variables:** Six general running variables are used: U<n>, V<n>, W<n>, X<n>, Y<n>, and Z<n>, as well as the dimensioned extensions.
- **Parameters:** Five parameters are employed: A<n>, B<n>, C<n>, D<n>, and E<n>, as well as their dimensioned extensions.

Further information regarding subroutine operation is given throughout the text.

I.5 Execution Speed

The program execution speed of an interpreter is slow by nature. However, as explained in Chapter VI, the math-intensive subroutines presented here are only a factor of four slower than analogous compiled routines. The variation in execution speed among the various BASICs is at least as great. For example, a fundamental comparison is shown in table I.1 (reference 41).

The timing comparisons shown for the subroutines given in this book correspond to standard eight-digit North Star BASIC. The fourteen-digit version runs about 40% slower. It is apparent that at least a two-fold increase in speed could be expected by using another BASIC. An additional two-fold improvement is possible by using a 4 MHz microprocessor instead of the 2 MHz one used for the comparisons given in the text. Thus, the values given in each chapter should be considered upper limits.

Execution Time In Milliseconds

Function	MITS 8 K (8080)	OSI 8 K (6502)	North Star (8080)	North Star + Floating-Point Board
multiply	4	2	5	2
divide	7	3	16	2
sine/cosine	23	17	99	11
logarithm	19	14	99	9
exponentiation	28	22	73	8

Table I.1 *A comparison of the execution speeds (in milliseconds) of some of the functions of various BASIC interpreters.*

CHAPTER II:

PLOTTING SUBROUTINES

One of the output capabilities often needed to technical and business computer programming is a method to graphically display data and functions. The next three sections describe software which semi-automatically plots given data and functions. These subroutines are designed to be compatible with most ACSII (American Standard Code for Information Interchange) terminals. They work best with hard copy devices since few video terminals can display as many lines as they can characters on a line, and the subroutines presented tend to print roughly square plots. To partially overcome the line number restriction with video terminals, the user may set the line width of the display so that the plot fits within the screen. (For further information on plotting routines, see reference 28.)

The first plotting routine presented handles equally spaced data values. The second routine treats the graphing of functions. It is a small variation on the equally spaced data program. The third routine is more complicated and displays coordinate-pair data points.

The plotting software is structured into three functional blocks. The first block, the calling program, is where data are created or organized. It may be part of a larger program, or, as shown below, it can be a simple data source. Using GOSUB, this program calls the actual plotting subroutine, which consists of two parts: a data-printing subroutine and an ordinate (Y) axis printer. This division is intentional. The same axis printing subroutine is used in all three examples considered. In two of the three examples, the same main plotting subroutine is employed. Thus, given in the three sections are three different calling programs, generally supplied by the user, two different main plotting subroutines, and only one axis printing subroutine. The structure is modular and allows the user considerable flexibility without undue complication.

The routines presented have several common visual output features. The software strives to produce a roughly square graph, with a tendency for the abscissa (X axis) to be longer than the ordinate (Y axis). The aspect ratio is adjustable, however, since the final appearance depends on the letter-to-letter and line-to-line spacing properties of the particular terminal being used.

The software also locates the Y=0 position and, if that value lies within the range of the data to be plotted, shifts the Y axis accordingly.

To maximize the plotting resolution of the terminal, the data are shifted and scaled such that the boundaries of the graph are determined by the extreme values of the data. As shown later, this feature can be bypassed if desired.

The execution time is largely limited by the program which feeds data to the plotting subroutine and the effective baud of the terminal. It is suggested that the line width specified for the plot be greater than 10 and less than 132.

In the three examples given in the following sections, we first look at the simplest case: plotting user-supplied data points which are equally spaced along the abscissa (X axis). In the second example, the same plotting subroutine is used to display data points generated by a function in the main calling program. Here, also, the output data are assumed to be equally spaced along the X axis. The third example treats the important case of plotting two-dimensional data sets. Since this type of data is usually not equally spaced in either the X or Y direction, a different plotting subroutine is introduced.

II.1 One-Dimensional Data Plot

As with all the plotting routines presented in this chapter, there is no great underlying mathematical basis to the technique. It is simply an exercise in the logical use of the PRINT and TAB functions. The routines are best understood by following the flow of the annotated programs shown.

The one-dimensional plotting subroutine is shown in program II.1b and starts (the first active subroutine statement) at line 40000. Program II.1a requests data from the terminal and calls the plotting subroutine to prepare the graph. The plotting subroutine requires the following specific information:

- L: the terminal line width to be used in plotting
- N: the number of data points to be displayed
- D(I): the N data points

The data matrix, D(I), must be put in ascending abscissa order by the calling program because the plotting program will simply graph the points in the order given. Although not intrinsically necessary, the beginning (X1) and ending (X2) abscissa values are also required by the calling program. They are not used in any calculations, but are printed as part of the display.

The functions and statements used, along with the variables list and parameters passed to the subroutine, are shown in table II.1. The program listing consists of three major program blocks. The associated memory requirements are shown in table II.2.

The first block is the calling program, which requests the data from the terminal. This is the program segment which could be replaced or modified by the user to meet a particular need. The main plotting subroutine starts at line 40000. The axis printing subroutine, which is called by the main plotting routine, actively starts at line 40200. Note that each subroutine is preceded by informative REMark statements, which are not vital to program execution. The subroutine connections are depicted in figure II.1.

Two program exercises are shown in listings II.1 and II.2. Two different data sets were plotted using terminal line widths of 50 and 32, respectively. Note how the zero position has been determined and the data scaled to give a full range plot.

One of the significant limitations of this plotting format is that the graph is at right angles to the normal sense (ie: the X axis is printed vertically). This is not a problem with hard copy listings, but is an annoyance with a video display. A longer subroutine given in section II.3 overcomes this difficulty.

It is possible to input data which cannot be properly plotted. If all the data values are above 100 million, the plot will not be scaled properly. This can be corrected by increasing the size of the dummy parameter, B, at the beginning of the subroutine, to the maximum value allowable in the BASIC being used.

As an extension of the use of the subroutine, the full-scale plotting feature can be artificially bypassed by inserting two dummy data values which represent the border values desired. For example, if all the data are between 3 and 7, yet a scale of 0 to 10 is desired, then two "data" values of 0 and 10 will accomplish that goal.

It is likely that the aspect ratio of the graph will not be optimum for the terminal used. This can be adjusted either by using the line-width parameter supplied by the calling program or by changing a number in statement 40037 (change 0.6 to a value more appropriate to the terminal being used). The value used can be determined experimentally.

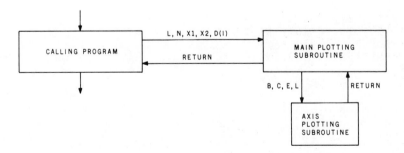

Figure II.1: *Subroutine connection for the data and function-plotting programs (EQAPLOT, AXISPLOT).*

Statements/Functions List

+, −, *, /, <, >
ABS, FOR/NEXT, GOSUB, GOTO, IF/THEN
INT, PRINT, TAB

Variables List

A, B, C, D(I), E, E2, E3, E4, I, J, K,
L, M, N, X1, X2

Variables Passed to Subroutine

D(I), L, N, X1, X2

Table II.1 *Statements, functions, and variables used by the plotting subroutine (EQAPLOT) shown in program II.1. In this and subsequent listings REM is not included since it is not necessary for operation, though the programs do contain many REMark statements; these can be removed to conserve memory. RETURN is also not included since it is understood to be used by every subroutine without exception.*

Program II.1:

(a)

```
PRINT"PROGRAM FOR PLOTTING EQUALLY"
PRINT"SPACED DATA VALUES"
PRINT
PRINT"THE USER INPUTS THE STARTING "
PRINT"AND ENDING COORDINATES, ALONG "
PRINT"WITH THE NUMBER OF EVALUATION POINTS."
PRINT"THE PROGRAM WILL THEN PLOT "
PRINT"THE DATA"
PRINT
PRINT
REM INITIALIZATION
PRINT "INPUT TERMINAL WIDTH: ",
INPUT L
PRINT "BEGINNING CORDINATE: ",
INPUT X1
PRINT "ENDING COORDINATE: ",
INPUT X2
PRINT "NUMBER OF DATA POINTS: ",
INPUT N
DIM D(N+1)
REM DATA INPUT
PRINT"INPUT DATA"
FOR I=1 TO N
PRINT I,
INPUT D(I)
NEXT I
REM GO TO PLOTTING SUBROUTINE PROPER
GOSUB 40000
END
```

Program II.1: (Cont.)

(b)

```
39998 REM PLOTTING SUBROUTINE (EQAPLOT)
39999 REM SHIFT DATA TO NON-NEGATIVE
40000 B=100000000
40001 REM FIND MINIMUM DATA VALUE
40002 FOR I=1 TO N
40003 IF B>D(I) THEN B=D(I)
40004 NEXT I
40005 REM SUBTRACT MINIMUM VALUE FROM ALL DATA
40006 FOR I=1 TO N
40007 D(I)=D(I)-B
40008 NEXT I
40009 REM FIND MAX. SHIFTED DATA VALUE
40010 C=0
40011 FOR I=1 TO N
40012 IF C<D(I) THEN C=D(I)
40013 NEXT I
40014 REM DETERMINE PRINTING SCALE VALUE
40015 A=L/C
40016 REM FIND TAB POSITION OF ZERO
40017 E=A*ABS(B)
40018 PRINT
40019 PRINT
40020 PRINT"***** DATA PLOT (SCALED) *****"
40021 PRINT
40022 PRINT
40023 PRINT"MIN. ORDINATE=   ",B,"    MAX. ORDINATE= ",C+B
40024 PRINT"INITIAL ABSCISSA VALUE= ",X1
40025 PRINT
40026 PRINT
40027 REM IF B IS POSITIVE, SKIP ZERO LABEL
40028 IF B>0 THEN GOTO 40034
40029 REM IF DATA ARE ALL BELOW ZERO, SKIP LABEL
40030 IF ABS(B)>C THEN GOTO 40034
40031 REM LABEL ZERO
40032 PRINT TAB(E),"0"
40033 REM GO TO AXIS PRINT SUBROUTINE
40034 GOSUB 40200
40035 FOR I=1 TO N
40036 REM INSERT LINE FEED FOR AUTO SPACING
40037 FOR K=1 TO (INT(0.6*L/N))
40038 PRINT":",TAB(L),":"
40039 NEXT K
40040 REM LOCATE DATUM POSITION
40041 E2=A*D(I)
40042 REM FORMATTED PRINT
40043 IF E2>=1 THEN GOTO 40046
40044 PRINT"*",
40045 GOTO 40049
40046 PRINT":",
40047 PRINT TAB(E2),"*",
40048 IF INT(E2)=L THEN GOTO 40050
40049 PRINT TAB(L),":",
40050 PRINT
40051 NEXT I
40052 REM GO TO AXIS PRINT SUBROUTINE
```

10 BASIC SCIENTIFIC SUBROUTINES

(b)
```
40053 GOSUB 40200
40054 PRINT
40055 PRINT
40056 PRINT"END ABSCISSA VALUE= ",X2
40057 PRINT
40058 PRINT
40059 REM RETURN TO DATA SOURCE PROGRAM
40060 RETURN
```

(c)
```
40199 REM AXIS PLOT (AXISPLOT)
40200 E3=E-5*INT(E/5)
40201 REM IF B IS POSTIVE, THEN SKIP ZERO LABEL
40202 IF B>0 THEN E3=0
40203 REM IF B IS GREATER THAN THE LARGEST VALUE, SKIP
40204 IF ABS(B)>C THEN E3=0
40205 FOR J=1 TO E3
40206 PRINT"-",
40207 NEXT J
40208 FOR J=1 TO (L-E3)/5
40209 PRINT"I----",
40210 NEXT J
40211 PRINT"I",
40212 E4=(J-1)*5+1+E3
40213 IF E4=L+1 THEN PRINT
40214 IF E4=L+1 THEN GOTO 40221
40215 E4=E4+1
40216 IF E4>=L+1 THEN GOTO 40219
40217 PRINT"-",
40218 GOTO 40215
40219 PRINT":"
40220 REM RETURN TO MAIN PLOTTING PROGRAM
40221  RETURN
```

Program II.1: *Program for plotting equally spaced data values. The plotting subroutine (EQAPLOT) actively starts at line 40000.*

Listing II.1:
```
RUN

PROGRAM FOR PLOTTING EQUALLY
SPACED DATA

THE USER INPUTS THE STARTING
AND ENDING COORDINATES, ALONG
WITH THE NUMBER OF DATA POINTS
THE PROGRAM WILL THEN PLOT
THE DATA.

INPUT TERMINAL WIDTH: ?50
BEGINNING CORDINATE: ?0
ENDING COORDINATE: ?12
NUMBER OF DATA POINTS: ?13
INPUT DATA
  1?-6
  2?-5
  3?-4
```

```
 4?-3
 5?-2
 6?-1
 7?0
 8?1
 9?2
10?3
11?4
12?5
13?6
```

***** DATA PLOT (SCALED) *****

```
MIN. ORDINATE=  -6   MAX. ORDINATE=  6
INITIAL ABSCISSA VALUE=  0
```

```
                         0
I----I----I----I----I----I----I----I----I----I
:*                                            :
:                                             :
:    *                                        :
:                                             :
:        *                                    :
:                                             :
:            *                                :
:                                             :
:                *                            :
:                                             :
:                    *                        :
:                                             :
:                        *                    :
:                                             :
:                            *                :
:                                             :
:                                *            :
:                                             :
:                                    *        :
:                                             :
:                                        *    :
:                                             :
:                                            *:
I----I----I----I----I----I----I----I----I----I
```

```
END ABSCISSA VALUE=   12
```

READY

Listing II.1: *Sample exercise using program II.1.*


```
RUN

PROGRAM FOR PLOTTING EQUALLY
SPACED DATA

THE USER INPUTS THE STARTING
AND ENDING COORDINATES, ALONG
WITH THE NUMBER OF DATA POINTS
THE PROGRAM WILL THEN PLOT
THE DATA.
```

Listing II.2:

```
INPUT TERMINAL WIDTH: ?32
BEGINNING CORDINATE: ?0
ENDING COORDINATE: ?14
NUMBER OF DATA POINTS: ?15
INPUT DATA
 1?-4
 2?-2
 3?1
 4?3
 5?5
 6?7
 7?5
 8?3
 9?1
 10?-1
 11?-3
 12?-3
 13?-2
 14?-1
 15?0


***** DATA PLOT (SCALED) *****

MIN. ORDINATE=   -4    MAX. ORDINATE=  7
INITIAL ABSCISSA VALUE=  0
```

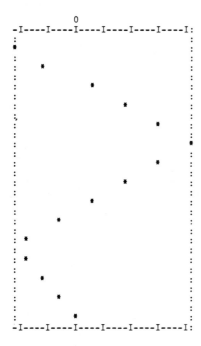

```
END ABSCISSA VALUE=   14

READY
```

Listing II.2: *Sample exercise using program II.1.*

Calling Program	565	bytes
Plotting Subroutine	1134	bytes
Axis Print Subroutine	415	bytes
TOTAL	2114	bytes

Table II.2 *Memory requirements for the programs appearing in section II.1.*

II.2 Function Plotting

The information in the previous section is applied below to create software which provides a graphical representation of a given function. The plotting subroutine discussed in the last section is used without alteration. However, the program that calls the subroutine is changed. This exemplifies how the plotting subroutine may be interfaced to a particular data source.

Program II.2 evaluates a given mathematical expression at N equally spaced points over a specified interval, (X1, X2). The expression to be evaluated is placed as indicated in the program. The input from the terminal is then L, the line width; X1 and X2, the interval; and N, the number of evaluation points. The calling program creates a data matrix and passes it to the subroutine.

(The listing shown in program II.2 does not contain the subroutine called since it has already been listed as program II.1b. It is implicitly assumed that subroutines 40000, the main plotting subroutine for equally spaced data—EQAPLOT, and 40200, the axis-printing subroutine—AXISPLOT, are part of the library. This referencing scheme is repeatedly used to save space, as well as to demonstrate the actual utility of a subroutine library. A subroutine cross-reference list is supplied in the appendices to aid in setting up the call locations.)

Most of the comments appearing in the previous section also apply to this section. The same functions and statements are used, with the possible exception of the function chosen by the user for evaluation. The calling program uses one additional variable, X. The only difference in the memory requirements comes from the change in the calling program shown in program II.2: 655 bytes, bringing the total to 2204 bytes.

Examples are shown in listings II.3 and II.4. Two different print-line widths were chosen to demonstrate how the plotting subroutine automatically adjusts the size of the presentation to keep it roughly square.

The utility of program II.2 can be enhanced by observing that a call to some other subroutine can replace the function expression appearing in the program.

```
PRINT"PROGRAM FOR PLOTTING EQUALLY"
PRINT"SPACED FUNCTION VALUES"
PRINT
PRINT"THE USER INPUTS THE STARTING "
PRINT"AND ENDING COORDINATES, ALONG "
PRINT"WITH THE NUMBER OF EVALUATION POINTS."
PRINT"THE PROGRAM WILL THEN PLOT "
PRINT"THE FUNCTION OVER THAT RANGE."
PRINT
PRINT
REM INITIALIZATION
PRINT "INPUT TERMINAL WIDTH: ",
INPUT L
PRINT "BEGINNING CORDINATE: ",
INPUT X1
PRINT "ENDING COORDINATE: ",
INPUT X2
PRINT "NUMBER OF DATA POINTS: ",
INPUT N
DIM D(N+1)
REM FUNCTION EVALUATION
FOR I=1 TO N
REM *****INPUT FUNCTION BELOW*****
X=X1+(I-1)*(X2-X1)/(N-1)
D(I)=.1*X*X*X-3*X*X+2*X-3
NEXT I
REM GO TO PLOTTING SUBROUTINE PROPER
GOSUB 40000
END
```

Program II.2: *Function plotting program example. Note that the called subroutines (EQAPLOT, AXISPLOT) are not listed. They are assumed to be part of the library. See program II.1.*

Listing II.3:
```
RUN

PROGRAM FOR PLOTTING EQUALLY
SPACED FUNCTION VALUES

THE USER INPUTS THE STARTING
AND ENDING COORDINATES, ALONG
WITH THE NUMBER OF EVALUATION POINTS.
THE PROGRAM WILL THEN PLOT
THE FUNCTION OVER THAT RANGE.


INPUT TERMINAL WIDTH: ?32
BEGINNING CORDINATE: ?0
ENDING COORDINATE: ?4
NUMBER OF DATA POINTS: ?12


***** DATA PLOT (SCALED) *****


MIN. ORDINATE=  -36.6    MAX. ORDINATE=  -2.664613
INITIAL ABSCISSA VALUE=   0
```

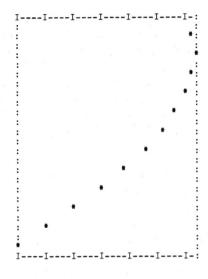

```
END ABSCISSA VALUE=   4

READY
```

Listing II.3: *Sample run using program II.2.*

```
RUN

PROGRAM FOR PLOTTING EQUALLY
SPACED FUNCTION VALUES

THE USER INPUTS THE STARTING
AND ENDING COORDINATES, ALONG
WITH THE NUMBER OF EVALUATION POINTS.
THE PROGRAM WILL THEN PLOT
THE FUNCTION OVER THAT RANGE.

INPUT TERMINAL WIDTH: ?64
BEGINNING CORDINATE: ?0
ENDING COORDINATE: ?4
NUMBER OF DATA POINTS: ?12


***** DATA PLOT (SCALED) *****


MIN. ORDINATE=    -36.6    MAX. ORDINATE=  -2.664613
INITIAL ABSCISSA VALUE=    0
```

Listing II.4:

16 BASIC SCIENTIFIC SUBROUTINES

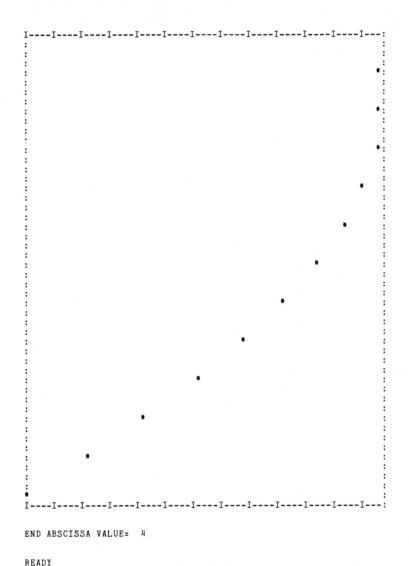

```
END ABSCISSA VALUE=   4

READY
```

Listing II.4: *Sample run using program II.2.*

II.3 Two-Dimensional Data Plot

The plotting routine presented in the previous sections is limited to the graphing of equally spaced data sets. Often experiments can be designed to give data in that form. However, in many cases the independent variable is not sampled in equal intervals, thus requiring a more general plotting subroutine.

The subroutine already presented may be modified to accomplish this by making the line spacing dependent on the abscissa interval size. Some additional logic must be included to handle the possibility of more than one data point at a given location or more than one data point on a given print line. In the plotting subroutine (40100; DATAPLOT) shown in program II.3b, the software recognizes any number of duplicate (triplicate, etc) points and prints only one symbol. Up to two different ordinate (Y) values may be printed on a given line (X). This limitation is not severe and permits plotting under most circumstances.

The calling program in program II.3a is required to present the (X,Y) coordinate pairs to the plotting subroutine in ascending or descending order, first according to abscissa (ascending or descending), and second according to ordinate (ascending). For instance, the coordinate pairs (1,1) and (2,0) are acceptable in sequence, but (1,1) and (1,0) are not. Also, among the coordinate-pair data set there must appear at least two different X values and two different Y values. Otherwise, the subroutine will attempt to make one of the axes zero in length, which is not permissible.

The use of this subroutine is demonstrated by the calling program shown in program II.3a. This program requests from the terminal the line width to be used, the number of coordinate points to be plotted, and the data set, one pair at a time.

The functions and statements employed are the same as those shown in section II.1. The variables list appears in table II.3.

Observe that, as before, L is the line width desired and N is the number of data points. C(I) and D(I) are the X and Y coordinate vectors. For example, (C(7), D(7)) represent the coordinates of the seventh data point.

Two examples are shown in listings II.5 and II.6. In the second example the coordinates are switched and the coordinate pairs rearranged according to the ascendency/descendency rules to produce a plot at right angles to the one before. The ability to rotate the graph is useful for video display presentation.

As indicated in section II.1, the aspect ratio for the graph depends on the terminal being used. This may be changed by adjusting the multiplying constant (0.5) in line 40118.

The memory requirements are shown in table II.4 and are similar to those given in section II.1.

The plotting subroutines presented in this chapter supply the user with a versatile means for graphical representation of data or functions.

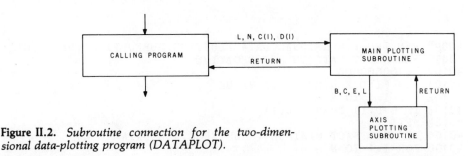

Figure II.2. *Subroutine connection for the two-dimensional data-plotting program (DATAPLOT).*

Variables List

A, B, C, C(I), D(I), E, E(I), E2, E3, E4,
I, J, K, L, N

Variables Passed to Subroutine

C(I), D(I), L, N

Table II.3 *Variables used by subroutine 40100, the two-dimensional data plot subroutine (DATAPLOT). See table II.1 for the functions and statements employed.*

Program II.3:
(a)
```
PRINT"PROGRAM FOR PLOTTING COORDINATE"
PRINT"SETS OF DATA"
PRINT
PRINT"THE PROGRAM ASSUMES THAT THE DATA IS"
PRINT"SEQUENTIALLY ORDERED, FIRST BY ABSCISSA"
PRINT"VALUE, SECOND BY ORDINATE"
PRINT"THE PROGRAM WILL THEN PLOT "
PRINT"THE DATA."
PRINT
PRINT
REM INITIALIZATION
PRINT "INPUT TERMINAL WIDTH: ",
INPUT L
PRINT "NUMBER OF DATA POINTS: ",
INPUT N
DIM D(N+1),C(N+1),E(N+1)
REM DATA INPUT
PRINT "INPUT DATA IN ABSCISSA, ORDINATE PAIRS: "
FOR I=1 TO N
PRINT I,
INPUT C(I),D(I)
REM C(I)=ABSCISSA, D(I)=ORDINATE
NEXT I
REM GO TO PLOTTING SUBROUTINE PROPER
GOSUB 40100
END
```

(b)
```
40098 REM TWO DIMENSIONAL DATA PLOTTING SUBROUTINE (DATAPLOT)
40099 REM SHIFT DATA TO NON-NEGATIVE
40100 B=100000000
40101 REM FIND MINIMUM DATA VALUE
40102 FOR I=1 TO N
40103 IF B>D(I) THEN B=D(I)
40104 NEXT I
40105 REM SUBTRACT MINIMUM VALUE FROM ALL DATA
40106 FOR I=1 TO N
```

Program II.3: (Cont.)
(b)

```
40107 D(I)=D(I)-B
40108 NEXT I
40109 REM FIND MAX. SHIFTED DATA VALUE
40110 C=0
40111 FOR I=1 TO N
40112 IF C<D(I) THEN C=D(I)
40113 NEXT I
40114 REM DETERMINE E(I), THE ABSCISSA SPACINGS
40115 E(0)=0
40116 E(N)=1
40117 FOR I=2 TO N
40118 E(I-1)=INT(0.5*(C(I)-C(I-1))*L/(C(N)-C(1))+.5)
40119 REM SPACING SCALED ACCORDING TO LINE WIDTH
40120 NEXT I
40121 REM DETERMINE PRINTING SCALE VALUE
40122 A=L/C
40123 REM FIND TAB POSITION OF ZERO
40124 E=A*ABS(B)
40125 PRINT
40126 PRINT
40127 PRINT"***** DATA PLOT (SCALED) *****"
40128 PRINT
40129 PRINT
40130 PRINT"MIN. ORDINATE=   ",B,"    MAX. ORDINATE= ",C+B
40131 PRINT"INITIAL ABSCISSA VALUE= ",C(1)
40132 PRINT
40133 PRINT
40134 REM IF B IS POSITIVE, SKIP ZERO LABEL
40135 IF B>0 THEN GOTO 40141
40136 REM IF DATA ARE ALL BELOW ZERO, SKIP LABEL
40137 IF ABS(B)>C THEN GOTO 40141
40138 REM LABEL ZERO
40139 PRINT TAB(E),"0"
40140 REM GO TO AXIS PRINT SUBROUTINE
40141 GOSUB 40200
40142 FOR I=1 TO N
40143 REM INSERT FEED FOR ABSCISSA SPACING
40144 FOR K=1 TO E(I-1)
40145 PRINT":",TAB(L),":"
40146 NEXT K
40147 REM LOCATE DATUM POSITION
40148 E2=A*D(I)
40149 REM TEST FOR MULTIPLE ORDINATE ABSCISSA
40150 IF E(I)=0 THEN GOTO 40161
40151 REM FORMATTED PRINT
40152 IF E2>=1 THEN GOTO 40155
40153 PRINT"*",
40154 GOTO 40158
40155 PRINT":",
40156 PRINT TAB(E2),"*",
40157 IF INT(E2)=L THEN GOTO 40159
40158 PRINT TAB(L),":",
40159 PRINT
40160 GOTO 40175
40161 REM TEST FOR OVERLAY OF POINTS
```

Program II.3:
(Cont.)
(b)

```
40162 REM IF DUPLICATE POINT, SKIP TO END
40163 IF D(I)=D(I+1) THEN GOTO 40175
40164 IF E2>=1 THEN GOTO 40168
40165 REM PRINT DOUBLE VALUES ON ONE LINE
40166 PRINT"*",TAB(A*D(I+1)),"*",
40167 GOTO 40171
40168 PRINT":",
40169 PRINT TAB(E2),"*",TAB(A*D(I+1)),"*",
40170 REM TEST IF LINE LIMIT REACHED
40171 IF INT(A*D(I+1))=L THEN GOTO 40173
40172 PRINT TAB(L),":",
40173 PRINT
40174 I=I+1
40175 NEXT I
40176 REM GO TO AXIS PRINT SUBROUTINE
40177 GOSUB 40200
40178 PRINT
40179 PRINT
40180 PRINT"END ABSCISSA VALUE= ",C(N)
40181 PRINT
40182 PRINT
40183 REM RETURN TO DATA SOURCE PROGRAM
40184 RETURN
```

Program II.3: *Demonstration program and subroutine (DATAPLOT) for plotting coordinate pair sets of data. Sample runs are shown on listings II.5 and II.6.*

Listing II.5:

```
RUN

PROGRAM FOR PLOTTING COORDINATE
SETS OF DATA

THE PROGRAM ASSUMES THAT THE DATA IS
SEQUENTIALLY ORDERED, FIRST BY ABSCISSA
VALUE, SECOND BY ORDINATE
THE PROGRAM WILL THEN PLOT
THE DATA.

INPUT TERMINAL WIDTH: ?64
NUMBER OF DATA POINTS: ?20
INPUT DATA IN ABSCISSA, ORDINATE PAIRS:
 1?18,15
 2?11,12
 3?10,9
 4?10,17
 5?9,7
 6?8,7
 7?8,8
 8?7,18
 9?6,2
 10?5,5
 11?5,5
 12?5,19
 13?4,2
 14?4,4
 15?2,21
 16?0,0
 17?-3,23
 18?-4,23
 19?-5,24
 20?-5,24
```

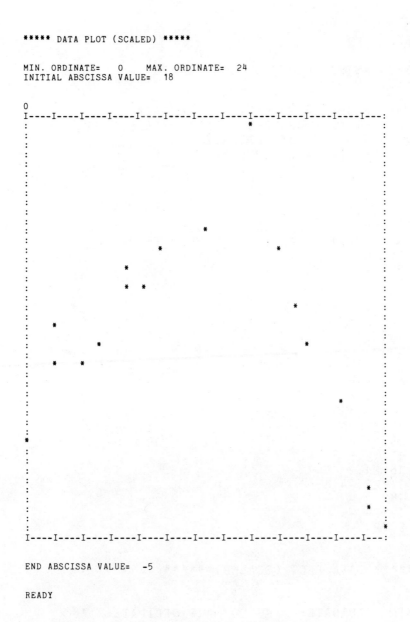

Listing II.5: *Two-dimensional data plot using program II.3.*

```
RUN

PROGRAM FOR PLOTTING COORDINATE
SETS OF DATA

THE PROGRAM ASSUMES THAT THE DATA IS
SEQUENTIALLY ORDERED, FIRST BY ABSCISSA
VALUE, SECOND BY ORDINATE
THE PROGRAM WILL THEN PLOT
THE DATA.


INPUT TERMINAL WIDTH: ?64
NUMBER OF DATA POINTS: ?20
INPUT DATA IN ABSCISSA, ORDINATE PAIRS:
 1?0,0
 2?2,4
 3?2,6
 4?4,4
 5?5,5
 6?5,5
 7?7,8
 8?7,9
 9?8,8
 10?9,10
 11?12,11
 12?15,18
 13?17,10
 14?18,7
 15?19,5
 16?21,2
 17?23,-4
 18?23,-3
 19?24,-5
 20?24,-5


***** DATA PLOT (SCALED) *****


MIN. ORDINATE=   -5    MAX. ORDINATE=   18
INITIAL ABSCISSA VALUE=   0
```

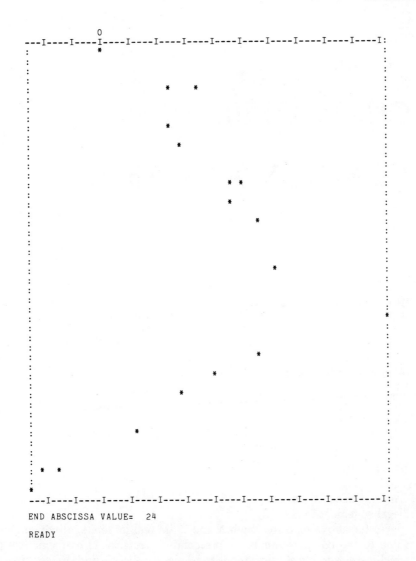

Listing II.6: *Two-dimensional data plot using program II.3. The (X,Y) coordinates used in listing II.5 were switched and rearranged to rotate the plot 90°.*

Calling Program:	571	bytes
Plotting Subroutine:	1689	bytes
Axis Subroutine:	415	bytes
TOTAL	2675	bytes

Table II.4 *Program memory size required for the two-dimensional data plot.*

CHAPTER III:

COMPLEX VARIABLES

III.1 The Complex Plane

One of the most curious abstractions in mathematics is the complex number. Complex numbers, which are composed of "real" and "imaginary" components, are often written in the form:

$$Z = X + iY \qquad (III.1)$$

where

i = the square root of -1

Care was taken not to define i as SQR(-1) or SQRT(-1) because that would be an illegal statement in BASIC.

In the above notation, both X and Y are real numbers, although Z is not. The value of the complex number representation rests in its mathematical properties which permit practical calculations to be made in simple and compact forms. For example, some functions are best integrated in the complex plane (see figure III.1) using the method of residues (see reference 27). Also, as will be explained shortly, complex numbers are intimately related to the sine and cosine functions and are thereby very useful in sinusoidal representations. A great deal of AC (alternating current) signal analysis is performed in the complex domain, with the results converted to meaningful descriptions of circuit performance, such as frequency response and phase shift.

In the following sections, subroutines are given for the addition, subtraction, multiplication, division, exponentiation and roots of complex numbers. This section deals with the two common coordinate representations of complex numbers, rectangular and polar.

As a vector, Z may be represented on an X and Y coordinate plane (see figure III.1). If a polar coordinate system, (U, V), is superimposed on that plane, Z can be equivalently described as the endpoint of a vector beginning at the origin, having length U and oriented at an angle V relative to the X axis. The relation between the (X,Y) and (U,V) coordinate systems is:

$$X = U \cos(V) \qquad \text{(III.2a)}$$

$$Y = U \sin(V) \qquad \text{(III.2b)}$$

$$U = (X^2 + Y^2)^{1/2} \qquad \text{(III.2c)}$$

$$V = \text{arctangent}(Y/X) \qquad \text{(III.2d)}$$

Figure III.1 is drawn with a circle to illustrate a very useful mathematical representation for Z:

$$Z = U \exp(iV) = U[\cos(V) + i \sin(V)] \qquad \text{(III.3)}$$

This form can be proven by using the Taylor series expansion of the exponential, observing the rules for powers of i (eg: the cube of i is $-i$), and collecting real and imaginary terms. Thus, $Z = X + iY$ can be obtained with the same coordinate transformation relationships as shown above. The exponential relationship will be extensively employed later, as it provides a simple means to multiply, divide, and perform other operations on complex numbers.

A subroutine for converting the (X,Y) representation of Z to the (U,V) representation is shown in program III.1. The conversion is identical to the transformation between rectangular (Cartesian) and polar coordinates, and therefore has utility beyond complex numbers. The manner in which the subroutine is connected to the calling program is shown in figure III.2. This subroutine may be exercised using program III.2, with the results shown in listing III.1. The size of these two programs and the execution speed of the subroutine are shown in tables III.1 and III.2. Information regarding the statements, functions, and variables used is given in table III.3. (This format is used extensively throughout the book.)

One particular property of this coordinate conversion subroutine is important in the interpretation of results. The angle returned (V) is in radians and has been adjusted to fall between 0 and 2π. This is consistent with the representation in figure III.1 and is assumed by the other subroutines given in this book. In some subroutine situations, inputting an angle outside this range will cause no difficulty, but the calling program should respect this restriction.

For the transformation of polar to rectangular coordinates, use the subroutine shown in program III.3. The manner in which this subroutine is connected to the calling program is detailed in figure III.3. A program for exercising this subroutine is described in program III.4, with sample results appearing in listing III.2. Observe in listing III.2 that some of the rectangular to polar conversions in listing III.1 were run backwards to demonstrate the transform/inverse transform properties of the two subroutines. Some error in the last digit displayed is apparent.

In the next section, subroutines are presented that perform other mathematical operations by calling the polar/rectangular coordinate conversion subroutines. Such use is the essence of the concept of a scientific subroutine library.

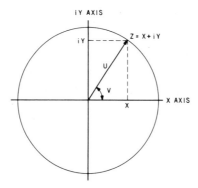

Figure III.1. *Polar representation of the complex-number plane.*

Subroutine	Bytes	Execution Time (milliseconds)
Add	89	42
Subtract	92	44
Rect/Polar	346	273
Polar/Rect	87	262
Polar Multiply	121	53
Polar Divide	106	56
Rect Multiply	330	901
Rect Divide	335	800
Polar Power	107	59
Polar Root	75	139
Rect Power	235	631
Rect Root	287	775
Spherical/Rect	130	550
Rect/Spherical	292	660

Table III.1 *Memory requirements and execution times for the subroutines given in Chapter III.*

Demonstration Program	Bytes
Add/Sub/Mult/Div	639
Rect/Polar Conversion	258
Polar/Rect Conversion	219
Complex No. to Power	288
Roots of a Complex No.	501
Spherical/Cartesian Conversion	262
Cartesian/Spherical Conversion	235

Table III.2 *Memory requirements for the non-subroutine parts of the demonstration programs given in Chapter III.*

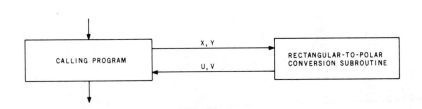

Figure III.2. *Subroutine connection for rectangular-to-polar conversion (RECT/POL).*

Statements/Functions List

$+, *, /, <, \wedge$
SQRT (or SQR), ATN (or ATAN), IF/THEN

Variables List

X, Y, U, V, W

Variables Passed to Subroutine

X, Y

Table III.3 *Statements, functions, and variables used in the rectangular-to-polar coordinate conversion subroutine (RECT/POL).*

```
40399 REM RECTANGULAR TO POLAR CONVERSION SUBROUTINE (RECT/POL)
40400 U=SQRT(X*X+Y*Y)
40401 REM GUARD AGAINST AMBIGUOUS VECTOR
40402 IF Y=0 THEN Y=(.1)^30
40403 REM GUARD AGAINST DIVIDE BY ZERO
40404 IF X=0 THEN X=(.1)^30
40405 REM SOME BASICS REQUIRE A SIMPLE ARGUMENT
40406 W=Y/X
40407 V=ATN(W)
40408 REM CHECK QUADRANT AND ADJUST
40409 IF X<0 THEN V=V+3.1415926535
40410 IF V<0 THEN V=V+6.2831853072
40411 RETURN
READY
```

Program III.1: *Rectangular coordinate-to-polar coordinate conversion subroutine (RECT/POL). See also figure III.2 and tables III.2 and III.3.*

```
REM PROGRAM TO DEMONSTRATE
REM RECTANGULAR TO POLAR CONVERSION
PRINT"INPUT RECTANGULAR COORDINATES TO BE CONVERTED",
INPUT X,Y
REM CONVERSION
GOSUB 40400
PRINT
PRINT"POLAR COORDINATES ARE : RADIUS= ",U
PRINT"                        ANGLE=  ",V," RADIANS"
END
```

Program III.2: *Program to demonstrate the use of the rectangular coordinate-to-polar coordinate conversion subroutine (RECT/POL).*

Listing III.1:
```
INPUT RECTANGULAR COORDINATES TO BE CONVERTED?1,0

POLAR COORDINATES ARE : RADIUS=   1
                        ANGLE=    9.9999999E-31 RADIANS




INPUT RECTANGULAR COORDINATES TO BE CONVERTED?-1,0

POLAR COORDINATES ARE : RADIUS=   1
                        ANGLE=    3.1415927 RADIANS
```

```
INPUT RECTANGULAR COORDINATES TO BE CONVERTED?0,2          Listing III.1: (Cont.)
POLAR COORDINATES ARE : RADIUS=   2
                        ANGLE=    1.5707963 RADIANS




INPUT RECTANGULAR COORDINATES TO BE CONVERTED?-3,-3
POLAR COORDINATES ARE : RADIUS=   4.2426407
                        ANGLE=    3.9269908 RADIANS
```

Listing III.1: *Sample results using program III.2, the rectangular-to-polar coordinate conversion (RECT/POL) demonstration program.*

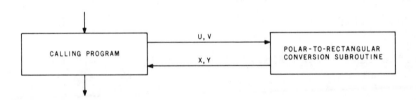

Figure III.3. *Subroutine connection for polar-to-rectangular coordinate conversion (POL/RECT).*

Statements/Functions List

*

COS
SIN

Variables List

X, Y, U, V

Variables Passed to Subroutine

U, V

Table III.4 *Statements, functions, and variables used in the polar-to-rectangular coordinate conversion subroutine (POL/RECT).*

```
40449 REM POLAR TO RECTANGULAR CONVERSION SUBROUTINE (POL/RECT)
40450 X=U*COS(V)
40451 Y=U*SIN(V)
40452 RETURN
```

Program III.3: *Polar coordinate-to-rectangular coordinate conversion subroutine (POL/RECT).*

```
REM PROGRAM TO DEMONSTRATE
REM POLAR TO RECTANGULAR CONVERSION
PRINT"INPUT POLAR COORDINATES TO BE CONVERTED",
INPUT U,V
REM CONVERSION
GOSUB 40450
PRINT
PRINT"RECTANGULAR COORDINATES ARE (X,Y)= ",X," ,",Y
END
```

Program III.4: *Program to demonstrate the use of the polar-to-rectangular coordinate conversion subroutine (POL/RECT) shown in program III.3. See listing III.2 for sample results.*

Listing III.2:

```
INPUT POLAR COORDINATES TO BE CONVERTED?1.4142136,.7853981
RECTANGULAR COORDINATES ARE (X,Y)=  1.0000001 , .99999997


INPUT POLAR COORDINATES TO BE CONVERTED?1,0
RECTANGULAR COORDINATES ARE (X,Y)=  1 , 0


INPUT POLAR COORDINATES TO BE CONVERTED?1,3.1415927
RECTANGULAR COORDINATES ARE (X,Y)=  -1 , -.0000001
```

```
INPUT POLAR COORDINATES TO BE CONVERTED?2,1.5707963
RECTANGULAR COORDINATES ARE (X,Y)=  0 , 2




INPUT POLAR COORDINATES TO BE CONVERTED?4.2426407,3.9269908
RECTANGULAR COORDINATES ARE (X,Y)=  -2.9999998 , -3.0000001
```

Listing III.2: (Cont.)

Listing III.2: *Sample runs of the polar-to-rectangular coordinate conversion subroutine (POL/RECT) using program III.4.*

III.2 Complex Variable Operations

In this section, the basic complex-number operations of addition, subtraction, multiplication, and division are considered. Subroutines for each of these four operations are given, followed by a calling program which exercises them.

Programs III.5 and III.6 show very simple subroutines for the addition and subtraction of complex numbers of the form $Z = X + iY$. Connection of these subroutines to the calling program appears in figure III.4. Information regarding their size and speed is presented in tables III.1 and III.2.

If the complex numbers to be added are in polar or exponential form (see section III.1), then the procedure is to convert them to rectangular form using the polar/rectangular transformation subroutine, and then perform the addition or subtraction. The result may be reconverted by using the rectangular/polar transformation subroutine.

The absolute value of a complex number is simply its length or radius, U. Thus, $|Z| = |X + iY|$ would be obtained by using the rectangular to polar conversion subroutine and extracting U.

To multiply two complex numbers in polar representation, the exponential form may be employed:

$$Z(1) = U(1) \exp[iV(1)] \tag{III.4a}$$

$$Z(2) = U(2) \exp[iV(2)] \tag{III.4b}$$

Thus,

$$Z = Z(1) \times Z(2) = U(1) \times U(2) \exp\{i[V(1) + V(2)]\} \tag{III.5}$$

The above operation is performed by the polar multiplication subroutine shown in program III.7. It may be incorporated into a rectangular complex-number multiplication subroutine by converting to polar coordinates, polar multiplying, and converting back. The subroutine in program III.8 performs this series of operations. Such a program is sometimes referred to as an executive, a program that directs the operation of other sub-programs.

Complex numbers in polar form can be divided very easily:

$$Z = \{U(1)/U(2)\} \exp\{i[V(1) - V(2)]\} \qquad (III.6)$$

A subroutine for doing this appears in program III.9. The corresponding executive subroutine for rectangular complex-number division is presented in program III.10. The scheme used for connecting the executive programs for rectangular complex-number multiplication and division to the calling program and the secondary subroutines to the executive is shown in figure III.5. The program size and execution speeds of the associated subroutines are detailed in table III.2.

A program for exercising several of the above subroutines is presented in program III.11. This program generated the examples appearing in listing III.3.

Several interesting properties of complex-number operations are apparent from an inspection of listing III.3. For example, the ratio of two complex numbers which are multiples of one another is simply that multiple. Also, the product of a complex-conjugate pair (eg: $X+iY$ and $X-iY$) is purely real.

The subroutines given in sections III.1 and III.2 assume that the variables transmitted to them are correct. No error detection is performed within the subroutines. For example, it is assumed that the radius of the input complex number (when in polar form) is positive and that the angle is between 0 and 2π. In addition, in the complex-number division subroutine, a zero divisor will cause a divide error. Note that either X or Y can be zero, but not simultaneously.

The subroutines provided in this section form the backbone of the program library for dealing with complex numbers. In the next section, several of these subroutines will be called upon to assist in performing the operations of raising complex numbers to a power and finding integer roots.

Figure III.4. *Subroutine connection for addition and subtraction of complex numbers (ZADD, ZSUB).*

Statements/Functions List

+ (or −)

Variables List

X(1), X(2), X(3), Y(1), Y(2), Y(3)

Variables Passed to Subroutine

X(1), X(2), Y(1), Y(2)

Table III.5 *Statements, functions, and variables used in the complex-number addition and subtraction subroutines (ZADD, ZSUB).*

```
40299 REM COMPLEX NUMBER ADDITION SUBROUTINE (ZADD)
40300 X(3)=X(1)+X(2)
40301 Y(3)=Y(1)+Y(2)
40302 RETURN
```

Program III.5: *Complex-number addition subroutine (ZADD).*

```
40349 REM COMPLEX NUMBER SUBTRACTION SUBROUTINE (ZSUB)
40350 X(3)=X(1)-X(2)
40351 Y(3)=Y(1)-Y(2)
40352 RETURN
```

Program III.6: *Complex-number subtraction subroutine (ZSUB).*

34 BASIC SCIENTIFIC SUBROUTINES

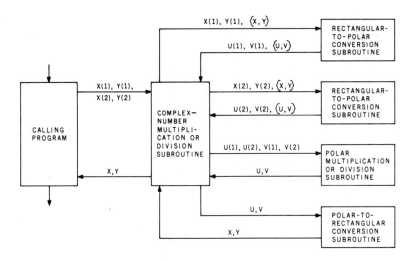

Figure III.5. *Subroutine connection for complex-number multiplication or division (ZRECTMLT, ZRECTDIV). Program variables actually passed to the subroutines are shown in brackets.*

Statements/Functions List

+, *, (or /), >
IF/THEN

Variables List

U(1), U(2), V(1), V(2), U, V

Variables Passed to Subroutine

U(1), U(2), V(1), V(2)

Table III.6 *Statements, functions, and variables used in the polar complex-number multiplication and division subroutines (ZPOLMLT, ZPOLDIV).*

Program III.7:

```
40499 REM POLAR MULTIPLICATION SUBROUTINE (ZPOLMLT)
40500 U=U(1)*U(2)
40501 V=V(1)+V(2)
40502 IF V>=6.2831853072 THEN V=V-6.2831853072
40503 RETURN
```

Program III.7: *Polar complex-number multiplication subroutine (ZPOLMLT).*

Statements/Functions List

GOSUB

Variables List

X(1), X(2), Y(1), Y(2)

Variables Passed to Subroutine

X(1), X(2), Y(1) and Y(2) are passed from the calling program to the executive subroutine.

Table III.7 *Statements, functions, and variables used in the rectangular complex-number multiplication and division subroutines (ZRECTMLT, ZRECTDIV). The items listed are in addition to those shown in tables III.3, III.4 and III.6.*

Program III.8:

```
40599 REM RECTANGULAR COMPLEX NUMBER MULTIPLICATION SUBROUTINE (ZRECTMLT)
40600 X=X(1)
40601 Y=Y(1)
40602 REM RECTANGULAR TO POLAR CONVERSION
40603 GOSUB 40400
40604 U(1)=U
40605 V(1)=V
40606 X=X(2)
40607 Y=Y(2)
40608 REM RECTANGULAR TO POLAR CONVERSION
40609 GOSUB 40400
40610 U(2)=U
40611 V(2)=V
40612 REM POLAR MULTIPLICATION
40613 GOSUB 40500
```

Program III.8:
(Cont.)

```
40614 REM POLAR TO RECTANGULAR CONVERSION
40615 GOSUB 40450
40616 RETURN
```

Program III.8: *Rectangular coordinates complex-number multiplication subroutine (ZRECTMLT). This program calls the rectangular-to-polar coordinate transformation subroutine (program III.1), its inverse (program III.3), and the polar multiplication subroutine (program III.7).*

```
40549 REM POLAR DIVISION SUBROUTINE (ZPOLDIV)
40550 U=U(1)/U(2)
40551 V=V(1)-V(2)
40552 IF V<0 THEN V=V+6.2831853072
40553 RETURN
```

Program III.9: *Polar coordinates complex-number division subroutine (ZPOLDIV).*

```
40799 REM RECTANGULAR COMPLEX NUMBER DIVISION SUBROUTINE (ZRECTDIV)
40800 X=X(1)
40801 Y=Y(1)
40802 REM RECTANGULAR TO POLAR CONVERSION
40803 GOSUB 40400
40804 U(1)=U
40805 V(1)=V
40806 X=X(2)
40807 Y=Y(2)
40808 REM RECTANGULAR TO POLAR CONVERSION
40809 GOSUB 40400
40810 U(2)=U
40811 V(2)=V
40812 REM POLAR COMPLEX NUMBER DIVISION
40813 GOSUB 40550
40814 REM POLAR TO RECTANGULAR CONVERSION
40815 GOSUB 40450
40816 RETURN
```

Program III.10: *Rectangular coordinates complex-number division subroutine (ZRECTDIV). The operation of this program is similar to the rectangular coordinates complex-number multiplication subroutine.*

```
REM PROGRAM TO DEMONSTRATE COMPLEX NUMBER
REM ADDITION, SUBTRACTION, MULTIPLICATION
REM AND DIVISION
PRINT"ENTER COMPLEX NUMBERS IN (X,Y) PAIRS:"
PRINT
PRINT"Z1= ",
INPUT X(1),Y(1)
PRINT
PRINT"Z2= ",
INPUT X(2),Y(2)
REM COMPLEX NUMBER ADDITION
GOSUB 40300
PRINT
PRINT
PRINT "Z1+Z2= ",X(3),
IF Y(3)>=0 THEN PRINT"+",
PRINT Y(3)," I"
PRINT
PRINT
REM COMPLEX NUMBER SUBTRACTION
GOSUB 40350
PRINT "Z1-Z2= ",X(3),
IF Y(3)>=0 THEN PRINT"+",
PRINT Y(3)," I"
PRINT
REM COMPLEX NUMBER MULTIPLICATION
GOSUB 40600
PRINT"Z1*Z2= ",X,
IF Y>=0 THEN PRINT" +",
PRINT Y," I"
PRINT
REM COMPLEX NUMBER DIVISION
GOSUB 40800
PRINT"Z1/Z2= X,
IF Y>=0 THEN PRINT" +",
PRINT Y," I"
END
```

Program III.11: *Program to demonstrate the use of the complex-number addition, subtraction, multiplication, and division subroutines (programs III.5, III.6, III.8, and III.10, respectively). Sample results are shown in listing III.3.*

```
ENTER COMPLEX NUMBERS IN (X,Y) PAIRS:

Z1= ?2,2

Z2= ?1,1


Z1+Z2=   3+  3 I

Z1-Z2=   1+  1 I

Z1*Z2=   4.0000001E-07 + 4.0000001 I

Z1/Z2=   1.9999999 + 0 I
```

Listing III.3:

Listing III.3:
(Cont.)

```
ENTER COMPLEX NUMBERS IN (X,Y) PAIRS:
Z1= ?3,-2
Z2= ?3,2

Z1+Z2=  6+ 0 I

Z1-Z2=  0 -4 I

Z1*Z2=  13 + 0 I

Z1/Z2=  .38461546 -.9230769 I



ENTER COMPLEX NUMBERS IN (X,Y) PAIRS:
Z1= ?2,1
Z2= ?-2,1

Z1+Z2=  0+ 2 I

Z1-Z2=  4+ 0 I

Z1*Z2=  -5.0000001 -5.0000001E-07 I

Z1/Z2=  -.59999999 -.79999999 I
```

Listing III.3: *Examples of complex-number addition, subtraction, multiplication, and division (ZADD, ZSUB, ZRECTMLT, ZRECTDIV) using program III.11.*

III.3 Powers and Roots of $Z = X + iY$

Taylor series expansions of analytic functions are not generally limited to real value arguments. In fact, such expansions necessarily have a region of convergence in the complex plane which includes the real-number subset. Since the Taylor series is simply a polynomial expansion in integral powers of the argument, the key to evaluating such series with complex arguments is to develop a way of finding the powers of complex numbers.

As in the case of complex-number multiplication and division, the operation is best accomplished using the exponential representation, in which case the Nth power of Z is:

$$Z = U^N \exp(iNV) \tag{III.7}$$

Refer to program III.12 for a subroutine that performs this polar operation. The integral power of a complex number in rectangular-coordinate representation may be computed by combining the polar-power subroutine with coordinate-conversion subroutines. The executive subroutine for coordinating this appears in program III.13 and is connected to the calling program as indicated in figure III.6.

The rectangular complex-number power subroutine may be exercised using program III.14, with sample results given in listing III.4. The first example is simply two to the fourth power (2^4), which is sixteen. Observe that the subroutine obtains this result by converting to polar coordinates, computing the power, and then converting back.

The second example, $2+2i$ to the sixth power (($2+2i)^6$), is counter-intuitive. In this case, we have a complex number with equal real and imaginary parts. Yet, when it is raised to the sixth power, the result is a purely imaginary number (it should be $-512i$). Note that the subroutine goes through the right motions in calculating the power, but has some error in it characteristic of computer inaccuracy.

The third example is simply the raising of i to the fifth power (i^5).

Finding the root of a complex number is very similar to raising a complex number to a power; 1 / N is used instead of N. However, the "phase" of the complex number has a cyclic property, based on the fact that $\exp(2M\pi i)=1$, where M is an integer. The complex number to be operated on could have an angle (in polar notation) of $V+2M\pi i$ without affecting its value in the $X+iY$ notation. Thus, there is some ambiguity (multiplicity) in the root:

$$Z^{1/N} = (U)^{1/N} \exp[i(V+2M\pi)/N] \qquad (III.8)$$

M can be any integer number. However, the roots repeat for M outside the range $0 < M < N-1$. Thus, there are N roots, and these are not necessarily distinct.

The subroutine shown in program III.15 calculates the lowest-order root when the complex number is in polar form. Program III.16 performs a similar operation for rectangular complex numbers, but returns with the Mth root. The scheme for accomplishing this is shown in figure III.7.

Program III.17 may be used to exercise the root-finding subroutine. It repeatedly calls the polar-root subroutine, each time obtaining a different root. These roots are tabulated in listing III.5.

The memory size and execution speed characteristics of the programs given in this section are indicated in the various tables. The rectangular complex-number power and root programs are slow. This is usually not a problem unless a great deal of number crunching is to be performed. A BASIC compiler would significantly improve the situation. Again, note that the subroutines assume the input to be valid. For example, if a negative value for U is transferred to the polar-root program, an argument error will result.

The subroutines presented here form a reasonably complete library for dealing with complex numbers. To aid in full utilization, these programs are cross-referenced in Appendix I according to both call line number and title. Compacted forms are also given in Appendix II.B; the subroutine call line numbers have been preserved. For further information, the reader is directed to references 12, 16, 17, 25, 26, and 27.

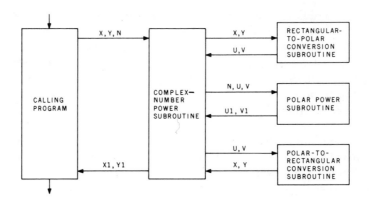

Figure III.6. *Subroutine connection for raising a complex number to an integer power (ZRECTPOW).*

Statements/Functions List

$+, -, *, /, \wedge$
INT

Variables List

N, U, U1, V, V1

Variables Passed to Subroutine

N, U, V

Table III.8 *Statements, functions, and variables used in the polar complex-number power subroutine (ZPOLPOW).*

```
41099 REM POLAR POWER SUBROUTINE (ZPOLPOW)
41100 U1=U^N
41101 V1=N*V
41102 V1=V1-6.2831853072*INT(V1/6.2831853072)
41103 RETURN
```

Program III.12: *Subroutine for raising a polar complex number to a power (ZPOLPOW).*

Statements/Functions List

GOSUB

Variables List

N, U, U1, V, V1

Variables Passed to Subroutine

N, X and Y are passed to the executive subroutine.

Table III.9 *Statements, functions, and variables used in the rectangular complex-number power subroutine (ZRECTPOW). Items listed are in addition to those shown in tables III.3, III.4 and III.8.*

```
41198 REM RECTANGULAR COMPLEX NUMBER POWER SUBROUTINE (ZRECTPOW)
41199 REM RECTANGULAR TO POLAR CONVERSION
41200 GOSUB 40400
41201 REM POLAR POWER
41202 GOSUB 41100
41203 REM CHANGE VARIABLE FOR CONVERSION
41204 U=U1
41205 V=V1
41206 REM POLAR TO RECTANGULAR CONVERSION
41207 GOSUB 40450
41208 RETURN
```

Program III.13: *Subroutine for raising a rectangular coordinate complex number to a power (ZRECTPOW). This program explicitly performs the required operation in polar representation (see program III.12) by first translating from rectangular to polar coordinates. The result of the polar power calculation is then transformed to rectangular coordinates. See figure III.6.*

```
REM PROGRAM TO DEMONSTRATE RAISING A
REM COMPLEX NUMBER TO A POWER
PRINT
PRINT"INPUT THE POWER DESIRED, FOLLOWED BY THE"
PRINT"COMPLEX NUMBER"
PRINT"N ",
INPUT N
PRINT"X ",
INPUT X
```

```
          PRINT"Y ",
          INPUT Y
          REM POWER CALCULATION
          GOSUB 41200
          PRINT
          PRINT"RESULT= ",X,
          IF Y1>=0 THEN PRINT" +",
          PRINT Y," I"
          END
```
Program III.14: (Cont.)

Program III.14: *Program for demonstrating the use of the rectangular complex-number power subroutine (ZRECTPOW). See listing 111.4 for sample results.*

```
          INPUT THE POWER DESIRED, FOLLOWED BY THE
          COMPLEX NUMBER
          N ?4
          X ?2
          Y ?0

          RESULT=  16 + 3.2E-29 I




          INPUT THE POWER DESIRED, FOLLOWED BY THE
          COMPLEX NUMBER
          N ?6
          X ?2
          Y ?2

          RESULT=  -1.5359999E-04 + -511.99998 I




          INPUT THE POWER DESIRED, FOLLOWED BY THE
          COMPLEX NUMBER
          N ?5
          X ?0
          Y ?-1

          RESULT=  .0000001 + -1 I
```
Listing III.4:

Listing III.4: *Sample results obtained using program III.14 (ZRECTPOW).*

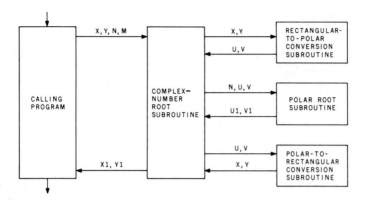

Figure III.7. *Subroutine connection for finding the Mth member of the Nth root of a complex number (ZRECTRT).*

Statements/Functions List

/ , ^

Variables List

N, U, U1, V, V1

Variables Passed to Subroutine

N, U, V

Table III.10 *Statements, functions, and variables used in the polar complex-number root subroutine (ZPOLRT).*

```
41149 REM POLAR (FIRST) ROOT SUBROUTINE (ZPOLRT)
41150 U1=U^(1/N)
41151 V1=V/N
41152 RETURN
```

Program III.15: *Subroutine for determining the Nth root of a complex number in polar representation (ZPOLRT).*

Statements/Functions List

GOSUB

Variables List

M

Variables Passed to Subroutine

M, N, X, and Y are passed to the executive subroutine.

Table III.11 *Statements, functions, and variables used in the rectangular complex-number root generation subroutine (ZRECTRT). The items listed are in addition to those appearing in tables III.4, III.5 and III.10.*

```
41298 REM RECTANGULAR COMPLEX NUMBER ROOT SUBROUTINE (ZRECTRT)
41299 REM RECTANGULAR TO POLAR CONVERSION
41300 GOSUB 40400
41301 REM POLAR (FIRST) ROOT
41302 GOSUB 41150
41303 U=U1
41304 REM FIND M ORDER ROOT
41305 REM M=1 CORRESPONDS TO THE FIRST ROOT
41306 V=V1+6.2831853072*(M-1)/N
41307 REM POLAR TO RECTANGULAR CONVERSION
41308 GOSUB 40450
41309 RETURN
```

Program III.16: *Subroutine for determining the Mth order root of a rectangular representation complex number (ZRECTRT). The polar root subroutine is called after performing a rectangular-to-polar coordinate conversion, and the result is transformed back to rectangular coordinates.*

Program III.17:
```
REM PROGRAM TO DEMONSTRATE
REM FINDING THE ROOTS OF A COMPLEX NUMBER
PRINT
PRINT"INPUT THE INTEGER ROOT DESIRED FOLLOWED"
PRINT"BY THE COMPLEX NUMBER"
PRINT"N ",
INPUT N
PRINT"X ",
REM X9 AND Y9 ARE STORED VALUES
```

Program III.17: (Cont.)

```
INPUT X9
PRINT"Y ",
INPUT Y9
REM FIND N ROOTS
PRINT
PRINT
PRINT"ORDER    X+ YI"
PRINT"-----   ----------------"
PRINT
FOR M=1 TO N
REM CONVERT TO X,Y
X=X9
Y=Y9
GOSUB 41300
PRINT"   ",M,
PRINT"   ",INT(1000*X+.5)/1000,"   ",
IF Y>=0 THEN PRINT "+",
PRINT INT(1000*Y+.5)/1000," I"
NEXT M
END
```

Program III.17: *Program to demonstrate the use of the rectangular coordinate representation complex-number root subroutine (program III.16). Sample results are shown in listing III.5.*

Listing III.5:

```
INPUT THE INTEGER ROOT DESIRED FOLLOWED
BY THE COMPLEX NUMBER
N ?2
X ?-1
Y ?0


ORDER    X+ YI
-----   ----------------

  1     0   + 1 I
  2     0    -1 I




INPUT THE INTEGER ROOT DESIRED FOLLOWED
BY THE COMPLEX NUMBER
N ?4
X ?1
Y ?0


ORDER    X+ YI
-----   ----------------

  1     1   + 0 I
  2     0   + 1 I
  3    -1     0 I
  4     0    -1 I
```

Listing III.5:
(Cont.)

```
INPUT THE INTEGER ROOT DESIRED FOLLOWED
BY THE COMPLEX NUMBER
N ?4
X ?0
Y ?1


ORDER    X+ YI
-----    ----------------

  1     .924  + .383 I
  2    -.383  + .924 I
  3    -.924  - .383 I
  4     .383  - .924 I
```

```
INPUT THE INTEGER ROOT DESIRED FOLLOWED
BY THE COMPLEX NUMBER
N ?8
X ?16
Y ?0


ORDER    X+ YI
-----    ----------------

  1    1.414  + 0 I
  2    1     + 1 I
  3    0     + 1.414 I
  4   -1     + 1 I
  5   -1.414 + 0 I
  6   -1     - 1 I
  7    0     - 1.414 I
  8    1     - 1 I
READY
```

```
INPUT THE INTEGER ROOT DESIRED FOLLOWED
BY THE COMPLEX NUMBER
N ?4
X ?16
Y ?-16


ORDER    X+ YI
-----    ----------------

  1     .425  + 2.139 I
  2   -2.139  +  .425 I
  3    -.425  - 2.139 I
  4    2.139  -  .425 I
READY
```

Listing III.5: *Sample runs using program III.17 to obtain the roots of complex numbers (ZRECTRT).*

III.4 Spherical Coordinate Conversion

The logical extension of conversions between rectangular and polar coordinates is to move up one dimension to spherical coordinates. The corresponding conversion programs are similar; in one case the subroutines given in a previous section are used.

The geometrical relationship between the spherical and Cartesian coordinate systems is indicated in figure III.8. The defining equations are:

$$X = U \sin(W) \cos(V) \qquad \text{(III.9a)}$$

$$Y = U \sin(W) \sin(V) \qquad \text{(III.9b)}$$

$$Z = U \cos(W) \qquad \text{(III.9c)}$$

When W, the angle off the Z axis, equals $\pi/2$, the equations for the X and Y coordinates reduce to the same ones used in the polar coordinate case. Program III.18 gives a simple subroutine for performing the spherical-to-rectangular (Cartesian) coordinate conversion calculations indicated. The schematic for connecting it to a calling program appears in figure III.9. Characteristics of the program are given in tables III.2 and III.12. Program III.19 may be used to apply this subroutine, with the sample results shown in listing III.6.

The reverse conversion, (X, Y, Z) to (U, V, W), is more complicated and is not often shown in software-oriented books. A subroutine for the transformation is presented in program III.20, with the corresponding connection diagram appearing in figure III.10. Application of this subroutine is shown with the aid of the demonstration routine given in program III.21, with examples displayed in listing III.7.

The coordinate conversion subroutines given in this chapter are simply based on the defining equations. It is possible to treat coordinate transformations as matrix operations. This will be considered for the following chapter.

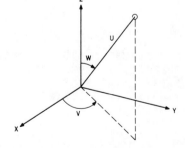

Figure III.8. *Spherical coordinate system.*

48 BASIC SCIENTIFIC SUBROUTINES

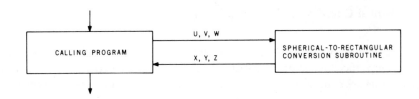

Figure III.9. *Subroutine connection for converting between spherical and rectangular coordinates (SPR/RECT).*

Statements/Functions List

*

SIN
COS

Variables List

X, Y, Z, U, V, W

Variables Passed to Subroutine

U, V, W

Table III.12 *Statements, functions, and variables used in the spherical-to-rectangular coordinate conversion subroutine (SPR/RECT).*

```
41399 REM SPHERICAL TO RECTANGULAR (CARTESIAN) CONVERSION
      SUBROUTINE (SPR/RECT)
41400 X=U*(SIN(W))*COS(V)
41401 Y=U*(SIN(W))*SIN(V)
41402 Z=U*COS(W)
41403 RETURN
READY
```

Program III.18: *Spherical-to-rectangular coordinate conversion subroutine (SPR/RECT).*

```
REM PROGRAM TO DEMONSTRATE SPHERICAL TO
REM CARTESIAN COORDINATE CONVERSION
PRINT"INPUT RADIUS, U: ",
INPUT U
PRINT"INPUT ANGLE IN (X,Y) PLANE, V: ",
INPUT V
PRINT"INPUT ANGLE OFF Z AXIS, W: ",
INPUT W
GOSUB 41400
PRINT
PRINT"(X,Y,Z)= (",X," , ",Y," , ",Z,")"
END
```

Program III.19: *Program to demonstrate spherical-to-rectangular coordinate conversion (SPR/RECT). See listing III.6.*

```
INPUT RADIUS, U: ?1
INPUT ANGLE IN (X,Y) PLANE, V: ?1
INPUT ANGLE OFF Z AXIS, W: ?1

(X,Y,Z)= ( .4546487 ,  .70807344 ,  .54030228)


INPUT RADIUS, U: ?2
INPUT ANGLE IN (X,Y) PLANE, V: ?0
INPUT ANGLE OFF Z AXIS, W: ?0

(X,Y,Z)= ( 0 ,  0 ,  2)


INPUT RADIUS, U: ?2
INPUT ANGLE IN (X,Y) PLANE, V: ?0
INPUT ANGLE OFF Z AXIS, W: ?1.57

(X,Y,Z)= ( 1.9999994 ,  0 ,  1.5925998E-03)


INPUT RADIUS, U: ?1
INPUT ANGLE IN (X,Y) PLANE, V: ?1.5707963268
INPUT ANGLE OFF Z AXIS, W: ?3.1415926536

(X,Y,Z)= ( 0 ,  -.0000001 ,  -1)
```

Listing III.6: *Sample spherical-to-rectangular coordinate conversions (SPR/RECT) using program III.19.*

50 BASIC SCIENTIFIC SUBROUTINES

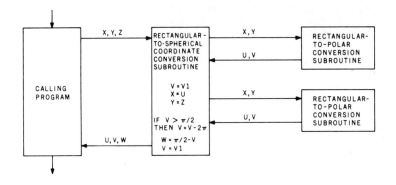

Figure III.10. *Subroutine connection for rectangular-to-spherical coordinate conversion (RECT/SPR).*

Statements/Functions List

−, >

GOSUB

Variables List

Z, W

Variables Passed to Subroutine

X, Y, and Z are passed to the executive subroutine.

Table III.13 *Statements, functions, and variables used in the rectangular-to-spherical coordinate conversion subroutine (RECT/SPR). Items listed are in addition to those in table III.4.*

Program III.20:
```
41448 REM RECTANGULAR (CARTESIAN) TO SPHERICAL CONVERSION
      SUBROUTINE (RECT/SPR)
41449 REM RECTANGULAR TO POLAR CONVERSION
41450 GOSUB 40400
41451 REM SAVE AND CHANGE VARIABLES
41452 V1=V
41453 X=U
41454 Y=Z
41455 REM RECTANGULAR TO POLAR CONVERSION
41456 GOSUB 40400
41457 IF V>1.5707963268 THEN V=V-6.28318553072
```

```
41458 W=1.5707963268-V
41459 V=V1
41460 RETURN
READY
```

Program III.20: *Rectangular-to-spherical coordinate conversion subroutine (RECT/SPR).*

Program III.20: (Cont.)

```
REM PROGRAM TO DEMONSTRATE CARTESIAN TO
REM SPHERICAL COORDINATE CONVERSION
PRINT"INPUT X, Y AND Z: ",
INPUT X,Y,Z
GOSUB 41450
PRINT
PRINT"RADIUS= ",U
PRINT"ANGLE IN (X,Y) PLANE= ",V," RADIANS"
PRINT"ANGLE OFF Z AXIS= ",W," RADIANS"
END
```

Program III.21: *Program to demonstrate rectangular-to-spherical coordinate conversion (RECT/SPR). See listing III.7.*

```
INPUT X, Y AND Z: ?.4546487,.70807344,.54030228

RADIUS= 1
ANGLE IN (X,Y) PLANE= 1 RADIANS
ANGLE OFF Z AXIS= 1 RADIANS




INPUT X, Y AND Z: ?0,0,2

RADIUS= 2
ANGLE IN (X,Y) PLANE= .7853981 RADIANS
ANGLE OFF Z AXIS= 0 RADIANS




INPUT X, Y AND Z: ?1.9999994,0,.0015925998

RADIUS= 2
ANGLE IN (X,Y) PLANE= 5.0000014E-31 RADIANS
ANGLE OFF Z AXIS= 1.57 RADIANS
```

Listing III.7:

```
INPUT X, Y AND Z: ?0,-.0000001,-1

RADIUS= 1
ANGLE IN (X,Y) PLANE=  4.712389 RADIANS
ANGLE OFF Z AXIS=  3.1415927 RADIANS
```

Listing III.7: *Sample rectangular-to-spherical coordinate conversions (RECT/SPR) using program III.21.*

CHAPTER IV

VECTOR AND MATRIX OPERATIONS

This chapter contains subroutines pertaining to vector and matrix algebra. Although the background discussion associated with these subroutines requires some familiarity with the general principles of algebra, the subroutines may be used without such knowledge to solve interesting problems, such as the angle between two vectors or simultaneous linear equations.

As in the previous chapters, information is supplied with each subroutine regarding its connection to the main program. It is assumed that the main program performs the task of dimensioning all the arrays used by the subroutines. The demonstration programs show how this may be done based on user input. Dimensioning is not performed in the subroutines themselves, since a given routine may be called more than once and a redimensioning error would result.

Note that the subroutines do not check the validity of the parameters passed to them. For example, the matrix inversion subroutine does not test the determinant of the matrix to be inverted for zero (a singularity), and thus the inversion routine will fail if it attempts to operate on such a matrix. The corresponding FORTRAN library subroutine would fail under similar conditions.

There are quite a few matrix operations for which subroutines might be written. The IBM SSP (*Scientific Subroutine Package*)(reference 29) was scanned to get an idea of the library content of that system. The following is a simplified abridgement of the matrix section of that library. None of the subroutines presented are BASIC translations of the FORTRAN programs given in reference 29. They are instead, reformulations based on the key text references, which are given in the associated discussion paragraphs. The main mathemetical reference for this entire chapter is *Advanced Engineering Mathematics* by Kreyszig (reference 27).

IV.1 Vector Operations

Vectors may be visualized as line segments which have both direction and length. A vector can be used to describe the velocity of an object in terms of both magnitude (speed) and direction. It can similarly be used to represent a force. Vectors have considerable application in mechanics and ballistics for the calculation of forces and displacement and in electromagnetics for the representation of electric and magnetic fields.

In three-dimensional rectangular (Cartesian) coordinates a vector **A** may be decomposed into three, orthogonal components:

$$\mathbf{A} = A(1)\ \hat{i} + A(2)\ \hat{j} + A(3)\ \hat{k} \tag{IV.1}$$

In this notation, $\hat{i}$, $\hat{j}$, and $\hat{k}$ are mutually orthogonal vectors of length 1. For the sake of algebraic manipulation, an equivalent form is the column vector:

$$\mathbf{A} = \begin{bmatrix} A(1) \\ A(2) \\ A(3) \end{bmatrix} \tag{IV.2}$$

The number of dimensions need not be limited to three, and all of the following vector-operation subroutines, except for the vector cross product, can be N-dimensional.

Vectors can be added, subtracted, and multiplied (in two different ways). It is also possible to determine the angle between two vectors. The following subroutines deal with these operations. Similar BASIC programs can be found in reference 17.

The sum of two three-dimensional vectors is defined as:

$$\mathbf{A} + \mathbf{B} = \begin{bmatrix} A(1) + B(1) \\ A(2) + B(2) \\ A(3) + B(3) \end{bmatrix} \tag{IV.3}$$

Vector subtraction is similarly defined. Subroutines 41500 (VECTADD) and 41550 (VECTSUB) perform these two operations for N-dimensional vectors (see programs IV.1 and IV.2). These subroutines are connected to the calling program as detailed in figure IV.1. Memory requirements and speed information for these programs are given in table IV.1.

Subroutine	Bytes	Time (North Star) (milliseconds)
Vector addition	98	50
Vector subtraction	101	50
Vector dot product	109	50
Vector cross product	153	50
Vector length	91	51
Vector angle	304	510
Matrix addition	124	225

Matrix subtraction	127	225
Matrix multiplication	228	960
Matrix transpose	132	165
Diagonal matrix generation	183	200
Matrix transfer (save)	392	200
Matrix scalar multiplication	398	220
Matrix clear	365	165
Matrix row switch	149	115
Matrix row add	137	80
Matrix cofactor	286	695
Matrix determinant	943	5400
Matrix inversion	1038	1900
Largest eigenvalue	766	800
Matrix exponent	914	2200/per series term evaluated

Table IV.1 *Memory requirements and execution times for the subroutines given in Chapter IV. In all but two cases, three-element vectors and 3-by-3 matrices were used for the timing evaluation. The two exceptions were the matrix cofactor (5 by 5 matrix) and determinant (4 by 4 matrix) subroutines. In all cases, the program length given is that required to simply load the subroutine. Additional space is automatically allocated when the vectors and matrices are dimensioned.*

Demonstration Program	Bytes
Vector add, subtract, cross product, dot product, length and angle	1480
Matrix printing utility subroutine	120
Matrix add, subtract, move, transpose, and multiply	3992
Matrix row switch and add, multiplication by scalar, and clearing	4029
Matrix determinants (to fourth rank)	2978
Matrix inversion	2877
Solving simultaneous equations	3961
Determining the largest eigenvalue	1506
Matrix exponentiation	3874

Table IV.2 *Memory size requirements to load the demonstration programs given in Chapter IV.*

Vectors can be multiplied in two ways. The first results in a scalar (the "dot" product), and the second results in a vector (the "cross" product). The dot product is defined in three dimensions to be:

$$\mathbf{A} \cdot \mathbf{B} = A(1) \times B(1) + A(2) \times B(2) + A(3) \times B(3) \tag{IV.4}$$

Program IV.3 gives a subroutine for performing this operation in N dimensions. See also references 12 and 25.

The vector cross product is more complicated, and although it may be generated for any N-dimensional space, it is usually applied only to the third dimension. In that case the definition is:

$$\mathbf{A} \times \mathbf{B} = \begin{bmatrix} A(2) \times B(3) - A(3) \times B(2) \\ A(3) \times B(1) - A(1) \times B(3) \\ A(1) \times B(2) - A(2) \times B(1) \end{bmatrix} \tag{IV.5}$$

See program IV.4 and references 12 and 25 for the vector cross-product subroutine.

Application of the vector cross product varies. It is very useful in calculating the force on a charged particle moving in a magnetic field (eg: a cyclotron) or a mass moving on the surface of a rotating sphere (eg: the wind on the earth's surface).

The length (or "norm") of a vector is simply defined as:

$$\text{length}(\mathbf{A}) = [A(1) \times A(1) + A(2) \times A(2) + A(3) \times A(3)]^{1/2} \tag{IV.6}$$

This definition can also be generalized to N dimensions. A subroutine for performing this simple operation appears in program IV.5.

The vector dot and cross products have simple geometrical interpretations. In the case of the dot product:

$$\mathbf{A} \cdot \mathbf{B} = \text{length}(\mathbf{A}) \times \text{length}(\mathbf{B}) \times \cos(E) \tag{IV.7}$$

E is the angle between the two multiplying vectors. For the cross product:

$$\mathbf{A} \times \mathbf{B} = \text{length}(\mathbf{A}) \times \text{length}(\mathbf{B}) \times \sin(E) \, \hat{n} \tag{IV.8}$$

$\hat{n}$ is a unit vector (length = 1) perpendicular to the plane containing the two multiplying vectors.

Either of the above two geometrical relations can be used to determine the angle between two vectors by using the subroutines given earlier. The shorter program for the dot product is used as follows:

$$\text{angle} = \arccos\{\mathbf{A} \cdot \mathbf{B} / [\text{length}(\mathbf{A}) \times \text{length}(\mathbf{B})]\} \tag{IV.9}$$

The subroutine appearing in program IV.6 performs this operation using the arctangent function found in many BASIC interpreters (ATN in North Star and ATAN in Microsoft). In Chapter VI, a subroutine is given for calculating the arctangent which allows this routine to be used with modification by BASICs not having an inverse trigonometric function.

The vector subroutines presented in this section may be demonstrated by use of program IV.7. Sample results are given in listing IV.1. In the first example **A** and **B** are in the (X, Z) plane, and therefore the cross product vector is in the Y direction, perpendicular to both multiplying vectors. In the second example **A** = **A**; the calculated angle between them is zero (with an accuracy of about .0003 radians). In the third example **A** = −**B**, and the subroutine correctly calculates the angle. The fourth example was chosen arbitrarily.

The subroutines in this section can also be applied to calculate the "direction cosines" of a vector. These are the cosines of the angles between the vector and the coordinate axes (X, Y, Z). If the vector is normalized (length(**A**) = 1), then the direction cosines are simply:

$$\cos(E1) = A(1)$$
$$\cos(E2) = A(2) \quad \text{(IV.10)}$$
$$\cos(E3) = A(3)$$

By virtue of the above relations and the fact that **A** is normalized, the sum of the squares of these cosines is 1. **A** may be normalized by dividing it by its length.

As a word of caution, remember to dimension the vectors in the calling program. Also, pass the right parameters!

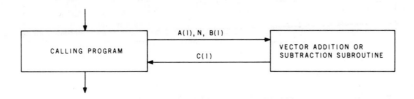

Figure IV.1. *Subroutine connection for vector addition or subtraction (VECTADD, VECTSUB).*

Statements/Functions List

+ (or −)

Variables List

A(I), B(I), C(I), N

Variables Passed to Subroutine

A(I), B(I), N

Table IV.3 *Statements, functions, and variables used in the vector addition and subtraction subroutines (VECTADD, VECTSUB).*

```
41498 REM VECTOR ADDITION SUBROUTINE (VECTADD)
41499 REM C=A+B
41500 FOR I=1 TO N
41501 C(I)=A(I)+B(I)
41502 NEXT I
41503 RETURN
```

Program IV.1: *Vector addition subroutine (VECTADD).*

```
41548 REM VECTOR SUBTRACTION SUBROUTINE (VECTSUB)
41549 REM C=A-B
41550 FOR I=1 TO N
41551 C(I)=A(I)-B(I)
41552 NEXT I
41553 RETURN
```

Program IV.2: *Vector subtraction subroutine (VECTSUB).*

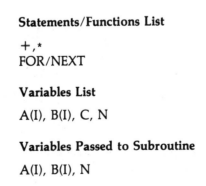

Figure IV.2. *Subroutine connection for the vector dot product (VECTDOT).*

Statements/Functions List

+, *
FOR/NEXT

Variables List

A(I), B(I), C, N

Variables Passed to Subroutine

A(I), B(I), N

Table IV.4 *Statements, functions, and variables used in the vector dot-product subroutine (VECTDOT).*

```
41598 REM VECTOR DOT PRODUCT SUBROUTINE (VECTDOT)
41599 REM C=A.B
41600 C=0
41601 FOR I=1 TO N
41602 C=C+A(I)*B(I)
41603 NEXT I
41604 RETURN
```

Program IV.3: *Vector dot-product subroutine (VECTDOT).*

Figure IV.3. *Subroutine connection for the vector cross product (VECTCURL).*

Statements/Functions List

−,*

Variables List

A(I), B(I), C(I)

Variables Passed to Subroutine

A(I), B(I)

Table IV.5 *Statements, functions, and variables used in the vector cross-product subroutine (VECTCURL).*

```
41648 REM VECTOR CROSS PRODUCT SUBROUTINE (VECTCURL)
41649 REM C=A X B
41650 C(1)=A(2)*B(3)-A(3)*B(2)
41651 C(2)=A(3)*B(1)-A(1)*B(3)
41652 C(3)=A(1)*B(2)-A(2)*B(1)
41653 RETURN
```

Program IV.4: *Vector cross-product subroutine (VECTCURL).*

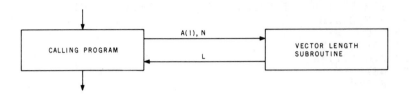

Figure IV.4. *Subroutine connection for determining the length of a vector (VECTLEN).*

Statements/Functions List

+,*

FOR/NEXT, SQRT (or SQR)

Variables List

A(I), L, N

Variables Passed to Subroutine

A(I), N

Table IV.6 *Statements, functions, and variables used in the vector length subroutine (VECTLEN).*

```
41699 REM VECTOR LENGTH SUBROUTINE (VECTLEN)
41700 L=0
41701 FOR I=1 TO N
41702 L=L+A(I)*A(I)
41703 NEXT I
41704 L=SQRT(L)
41705 RETURN
```

Program IV.5: *Vector length subroutine (VECTLEN).*

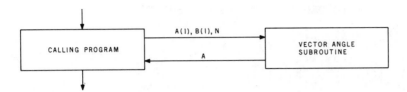

Figure IV.5. *Subroutine connection for determining the angle between two vectors (VECTANGL).*

Statements/Functions List

$+, -, *, /, <,$
ATN (or ATAN), FOR/NEXT, GOSUB, IF/THEN

Variables List

A, A(I), B(I), C, E, L, L1, N

Variables Passed to Subroutine

A(I), B(I), N

Table IV.7 *Statements, functions, and variables used in the vector angle subroutine (VECTANGL). Also requires subroutines 41600 and 41700 (see Appendix II).*

```
41747 REM VECTOR ANGLE SUBROUTINE (VECTANGL)
41748 REM ANGLE BETWEEN A AND B
41749 REM FIND DOT PRODUCT
41750 GOSUB 41600
41751 REM FIND LENGTH OF A
41752 GOSUB 41700
41753 REM SAVE VALUE
41754 L1=L
41755 REM FIND LENGTH OF B
41756 FOR I=1 TO N
41757 A(I)=B(I)
41758 NEXT I
41759 GOSUB 41700
41760 E=C/(L*L1)+(.1)^30
41761 E=SQRT(1-E*E)/E
41762 A=ATN(E)
41763 IF C<0 THEN A=3.1415926536-A
41764 RETURN
```

Program IV.6: *Subroutine to determine the angle between two vectors (VECTANGL).*

```
REM PROGRAM TO DEMONSTRATE VECTOR SUBROUTINES
N=3
DIM A(N), B(N), C(N)
PRINT "INPUT VECTOR A (X,Y,Z): ",
INPUT A(1),A(2),A(3)
PRINT "INPUT VECTOR B (X,Y,Z): ",
INPUT B(1),B(2),B(3)
REM SUM
GOSUB 41500
PRINT
PRINT "A+B= (",C(1),", ",C(2),", ",C(3),")"
REM DIFFERENCE
GOSUB 41550
PRINT
PRINT"A-B= (",C(1),", ",C(2),", ",C(3),")"
REM CROSS PRODUCT
GOSUB 41650
PRINT
PRINT"AXB= (",C(1),", ",C(2),", ",C(3),")"
REM DOT PRODUCT
GOSUB 41600
PRINT
PRINT "A.B= ",C
REM VECTOR ANGLE
GOSUB 41750
PRINT
PRINT "ANGLE BETWEEN VECTORS A AND B= ",A," RADIANS"
REM VECTOR LENGTH
GOSUB 41700
PRINT
PRINT "LENGTH OF VECTOR A= ",L
PRINT
END
```

Program IV.7: *Program to demonstrate the application of the vector operation subroutines (programs IV.1 thru IV.6).*

Listing IV.1:
```
INPUT VECTOR A (X,Y,Z): ?1,0,0
INPUT VECTOR B (X,Y,Z): ?0,0,1

A+B= ( 1,   0,   1)

A-B= ( 1,   0,  -1)

AXB= ( 0,  -1,   0)

A.B=  0

ANGLE BETWEEN VECTORS A AND B=  1.5707963 RADIANS

LENGTH OF VECTOR A=  1
```

```
INPUT VECTOR A (X,Y,Z): ?1,1,0
INPUT VECTOR B (X,Y,Z): ?1,1,0

A+B= ( 2,  2,  0)

A-B= ( 0,  0,  0)

AXB= ( 0,  0,  0)

A.B=  2

ANGLE BETWEEN VECTORS A AND B=  3.1622777E-04 RADIANS

LENGTH OF VECTOR A=  1.4142136




INPUT VECTOR A (X,Y,Z): ?1,0,0
INPUT VECTOR B (X,Y,Z): ?-1,0,0

A+B= ( 0,  0,  0)

A-B= ( 2,  0,  0)

AXB= ( 0,  0,  0)

A.B=  -1

ANGLE BETWEEN VECTORS A AND B=  3.1415927 RADIANS

LENGTH OF VECTOR A=  1




INPUT VECTOR A (X,Y,Z): ?1,1,1
INPUT VECTOR B (X,Y,Z): ?0,1,0

A+B= ( 1,  2,  1)

A-B= ( 1,  0,  1)

AXB= ( -1,  0,  1)

A.B=  1

ANGLE BETWEEN VECTORS A AND B=  .9553166 RADIANS

LENGTH OF VECTOR A=  1
```

Listing IV.1: *Sample runs of vector-operation subroutines using program IV.7.*

IV.2 Matrix Sums and Products

The remainder of this chapter deals with matrix operations. Matrices are simply orderly arrays of numbers. Vectors are one-dimensional matrices consisting of a single column. Two-dimensional matrices have the form:

$$A = \begin{bmatrix} A(1,1) & A(1,2) \\ A(2,1) & A(2,2) \\ A(3,1) & A(3,2) \end{bmatrix} \qquad (IV.11)$$

where A is a three-row by two-column matrix.

Matrix algebra provides a set of very powerful techniques for handling arrays. These arrays may represent data which is to be transformed into a more easily understood structure. Experimenters and engineers may find this section particularly useful. For example, the experimenter may have a large set of data relating a collection of input variables to a group of responses and may wish to determine the form of the (assumed) linear relationships using matrix-oriented statistical techniques. Alternatively, an engineer may have a mechanical system which can be described as a combination of springs and masses and may wish to determine the modes of vibration. In a similar vein, an electrical engineer may have sketched out a feedback control system and may wish to determine the stability of the system. In the latter two cases, a matrix which represents the coefficients of a series of equations is processed to determine the *eigenvalues* (see reference 27); these contain the answer to the problem.

In the following pages, the fundamental subroutines required to perform many such analyses will be presented along with interesting examples. These subroutines will permit the programmer to deal with matrix operations symbolically, leaving the drudgery to the computer.

As with vectors, matrices can be added or subtracted. Each element of the resulting matrix is simply the sum or difference of the corresponding elements in the two matrices being acted upon. If I denotes the row number and J the column number of each M by N matrix:

Sum: $C(I,J) = A(I,J) + B(I,J)$
Difference: $C(I,J) = A(I,J) - B(I,J)$

Observe that $I \leq M$ and $J \leq N$.

The operation must be performed for each element, and there are $M \times N$ of them. Subroutines for these two operations are shown in programs IV.8 and IV.9. These subroutines are connected to the calling program as indicated in figure IV.6, with other relevant information given in tables IV.1 and IV.8.

Note that the vector addition and subtraction subroutines are redundant with respect to the matrix subroutines in that an N-dimensional vector can be represented by a matrix having N rows and only one column. Vector addition and subtraction may thus be performed using the corresponding matrix addition and subtraction subroutines. This equivalency will be used later in both solving simultaneous equations and obtaining eigenvalues. Refer to references 17 and 22 for some corresponding programs in BASIC and FORTRAN, respectively.

Matrices can also be multiplied. This is an important operation because the majority of matrix transformations are performed in this manner. The matrix product is similar to the vector dot product, but on an element-by-element basis. For element C(I,J) of the product matrix:

$$C(I,J) = A(I,1) \times B(1,J) + A(I,2) \times B(2,J) + \cdots$$

or, in more concise form:

$$C(I,J) = \sum_{K=1}^{N} A(I,K) \times B(K,J) \tag{IV.12}$$

Observe that the above definition requires that the number of columns in matrix A must equal the number of rows in matrix B. The subroutine in program IV.10 performs this calculation and assumes the column/row equality. Note that matrix multiplication requires the calculation of quite a few products, thus making the operation slow. See table IV.1 for execution time information.

From the above definitions it is apparent that the order of calculation is not important for the matrix sum and difference operations. That is,

$$A + B = B + A$$
$$A - B = -B + A$$

However, the order for multiplication is very important. In general,

$$A \times B \neq B \times A$$

In fact, the two multiplications can only be performed in either order only if the matrices are *square* (ie: the number of rows equals the number of columns). The user must supply the proper matrices to the subroutine.

One of the ways in which the order of multiplication can be switched is to *transpose* the multiplying matrices. Thus, if the product A × B is legal, then so is the product (transpose B) × (transpose A). The transpose operation simply involves the switching of rows and columns. Row 1 becomes column 1; row 2 becomes column 2, and so on. Such a function is provided by the subroutine given in program IV.11. Again, the calling program must pass the appropriate information to the subroutine.

The transpose is a fundamental operation. The next few subroutines provide useful matrix algebra utility functions. The first creates a diagonal matrix with each diagonal element having a value B (see program IV.12). A diagonal matrix is necessarily square. The 3 by 3 case appears below:

$$\text{Matrix B} = \begin{bmatrix} B & 0 & 0 \\ 0 & B & 0 \\ 0 & 0 & B \end{bmatrix} = B \times \begin{bmatrix} 1 & 0 & 0 \\ 0 & 1 & 0 \\ 0 & 0 & 1 \end{bmatrix} = B \times I \tag{IV.13}$$

The above equation introduces the identity matrix I, a diagonal matrix having all 1s on the diagonal. By passing B = 1 to the subroutine, the identity matrix can be generated. It is given that name because when it multiplies a matrix (assuming the number of columns in the identity matrix equals the number of rows in the matrix being multiplied), the result is the matrix multiplied. That is,

$$I \times B = B \tag{IV.14a}$$

Similarly,

$$B \times I = B \tag{IV.14b}$$

The identity matrix will be encountered again in section IV.8, where the subject of eigenvalues is considered.

The next set of utility programs permits the saving of one matrix in another matrix in six combinations (see programs IV.13 thru IV.18). As indicated in figure IV.10, there are three source matrices and three destination matrices: A, B, and C. These matrices can be one-, two-, or three-dimensional as determined by the values of N1, N2, and N3. For example, if N2 and N3 are both zero, then the subroutine will treat the source and destination matrices as vectors (one-dimensional matrices) having N1 elements. If N3 is zero, then N1 (row) by N2 (column) matrices are assumed. In most cases, N3 = 0 will be chosen. It is a default option in that if N3 is not used in the calling program, it will be assigned a value of zero upon initialization. The matrix-save operation is used extensively in later programs.

The matrix operation subroutines presented thus far may be exercised by program IV.19, with sample results appearing in listings IV.2a and IV.2b. Program IV.20 shows a matrix-printing subroutine called by the demonstration program. It is not part of the general subroutine package, and therefore its statement numbers will not be found in the 40000 thru 50000 block reserved for the library. The user will probably wish to design his or her own printing format.

The sample runs demonstrate some interesting matrix properties. For example, in listing IV.2a, observe that when a three-row by two-column matrix B is multiplied by a two-row by three-column matrix (transpose A), the result is a two-row by two-column matrix. In listing IV.2b, it is apparent that the transpose of a diagonal matrix equals itself. Also, multiplication by the identity matrix does not affect the multiplied matrix.

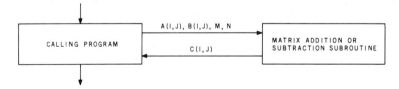

Figure IV.6. *Subroutine connection for matrix addition or subtraction (MATADD, MATSUB).*

Statements/Functions List

+ (or −)
FOR/NEXT

Variables List

A(I,J), B(I,J), C(I,J), M, N

Variables Passed to Subroutine

A(I,J), B(I,J), M, N

Table IV.8 *Statements, functions, and variables used in the matrix addition and subtraction subroutines (MATADD, MATSUB).*

```
41798 REM MATRIX ADDITION SUBROUTINE (MATADD)
41799 REM C=A+B
41800 FOR I=1 TO M
41801 FOR J=1 TO N
41802 C(I,J)=A(I,J)+B(I,J)
41803 NEXT J
41804 NEXT I
41805 RETURN
```

Program IV.8: *Matrix addition subroutine (MATADD).*

```
41848 REM MATRIX SUBTRACTION SUBROUTINE (MATSUB)
41849 REM C=A-B
41850 FOR I=1 TO M
41851 FOR J=1 TO N
41852 C(I,J)=A(I,J)-B(I,J)
41853 NEXT J
41854 NEXT I
41855 RETURN
```

Program IV.9: *Matrix subtraction subroutine (MATSUB).*

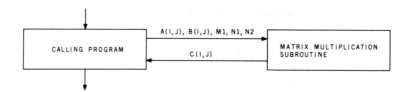

Figure IV.7. *Subroutine connection for matrix multiplication (MATMULT).*

Statements/Functions List

+,*

FOR/NEXT

Variables List

A(I,K), B(K,J), C(I,J), M1, N1, N2

Variables Passed to Subroutine

A(I,K), B(K,J), M1, N1, N2

Table IV.9 *Statements, functions, and variables used in the matrix multiplication subroutine (MATMULT).*

```
41898 REM MATRIX MULTIPLICATION SUBROUTINE (MATMULT)
41899 REM C=A X B    A IS M1 BY N1    B IS M2 BY N2    C IS M1 BY N2
41900 FOR I=1 TO M1
41901 FOR J=1 TO N2
41902 C(I,J)=0
41903 FOR K=1 TO N1
41904 C(I,J)=C(I,J)+A(I,K)*B(K,J)
41905 NEXT K
41906 NEXT J
41907 NEXT I
41908 RETURN
41948 REM MATRIX TRANSPOSE SUBROUTINE (MATTRANS)
```

Program IV.10: *Matrix multiplication subroutine (MATMULT).*

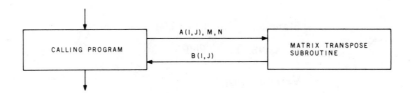

Figure IV.8. *Subroutine connection for transposing a matrix (MATTRANS).*

Statements/Functions List
FOR/NEXT

Variables List
A(I,J), B(I,J), M, N

Variables Passed to Subroutine
A(I,J), M, N

Table IV.10 *Statements, functions, and variables used in the matrix transpose subroutine (MATTRANS).*

```
41948 REM MATRIX TRANSPOSE SUBROUTINE (MATTRANS)
41949 REM B=TRANSPOSE(A)
41950 FOR I=1 TO N
41951 FOR J=1 TO M
41952 B(I,J)=A(J,I)
41953 NEXT J
41954 NEXT I
41955 RETURN
```

Program IV.11: *Matrix transpose subroutine (MATTRANS).*

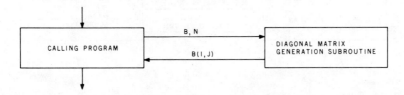

Figure IV.9. *Subroutine connection for creating a diagonal matrix (MATDIAG).*

Statements/Functions List

FOR/NEXT, IF/THEN

Variables List

B, B(I,J), N

Variables Passed to Subroutine

B, N

Table IV.11 *Statements, functions, and variables used in the diagonal-matrix generation subroutine (MATDIAG).*

```
41998 REM DIAGONAL MATRIX CREATION SUBROUTINE (MATDIAG)
41999 REM MATRIX B(I,J) IS THE IDENTITY MATRIX TIMES B
42000 FOR I=1 TO N
42001 FOR J=1 TO N
42002 B(I,J)=0
42003 IF I=J THEN B(I,J)=B
42004 NEXT J
42005 NEXT I
42006 RETURN
```

Program IV.12: *Diagonal-matrix creation subroutine (MATDIAG).*

Figure IV.10. *Subroutine connection for saving a source matrix (A, B, C) in a destination matrix (A, B, C). Six combinations are possible.*

Statements/Functions List

*

IF/THEN, FOR/NEXT

Variables List

Two matrices: A( ), B( ), or C( )
I1, I2, I3, N1, N2, N3

Variables Passed to Subroutine

One matrix: one of A( ), B( ), or C( )
N1, N2, N3

Table IV.12 *Statements, functions, and variables used in the matrix save subroutines. The matrices involved may be of one, two, or three dimensions as indicated by N1, N2, and N3.*

```
42048 REM MATRIX SAVE (A IN B) SUBROUTINE (MATSAVAB)
42049 REM N1,N2 AND N3 ARE INPUT INDICES
42050 IF N1*N2*N3=0 THEN GOTO 42060
42051 REM CHECK DIMENSION
42052 FOR I1=1 TO N1
42053 FOR I2=1 TO N2
42054 FOR I3=1 TO N3
42055 B(I1,I2,I3)=A(I1,I2,I3)
42056 NEXT I3
42057 NEXT I2
42058 NEXT I1
42059 RETURN
42060 IF N1*N2=0 THEN GOTO 42067
42061 FOR I1=1 TO N1
42062 FOR I2=1 TO N2
42063 B(I1,I2)=A(I1,I2)
42064 NEXT I2
42065 NEXT I1
42066 RETURN
42067 IF N1=0 THEN RETURN
42068 FOR I1=1 TO N1
42069 B(I1)=A(I1)
42070 NEXT I1
42071 RETURN
```

Program IV.13: *Matrix save subroutine (MATSAVAB). Matrix A is saved in matrix B.*

```
42073 REM MATRIX SAVE (B IN A) SUBROUTINE (MATSAVBA)
42074 REM N1,N2 AND N3 ARE INPUT INDICES
42075 IF N1*N2*N3=0 THEN GOTO 42085
```

Program IV.14:

```
42076 REM CHECK DIMENSION
42077 FOR I1=1 TO N1
42078 FOR I2=1 TO N2
42079 FOR I3=1 TO N3
42080 A(I1,I2,I3)=B(I1,I2,I3)
42081 NEXT I3
42082 NEXT I2
42083 NEXT I1
42084 RETURN
42085 IF N1*N2=0 THEN GOTO 42092
42086 FOR I1=1 TO N1
42087 FOR I2=1 TO N2
42088 A(I1,I2)=B(I1,I2)
42089 NEXT I2
42090 NEXT I1
42091 RETURN
42092 IF N1=0 THEN RETURN
42093 FOR I1=1 TO N1
42094 A(I1)=B(I1)
42095 NEXT I1
42096 RETURN
```

Program IV.14: *Matrix save subroutine (MATSAVBA). Matrix B is saved in matrix A.*

```
42098 REM MATRIX SAVE (C IN B) SUBROUTINE (MATSAVCB)
42099 REM N1,N2 AND N3 ARE INPUT INDICES
42100 IF N1*N2*N3=0 THEN GOTO 42110
42101 REM CHECK DIMENSION
42102 FOR I1=1 TO N1
42103 FOR I2=1 TO N2
42104 FOR I3=1 TO N3
42105 B(I1,I2,I3)=C(I1,I2,I3)
42106 NEXT I3
42107 NEXT I2
42108 NEXT I1
42109 RETURN
42110 IF N1*N2=0 THEN GOTO 42117
42111 FOR I1=1 TO N1
42112 FOR I2=1 TO N2
42113 B(I1,I2)=C(I1,I2)
42114 NEXT I2
42115 NEXT I1
42116 RETURN
42117 IF N1=0 THEN RETURN
42118 FOR I1=1 TO N1
42119 B(I1)=C(I1)
42120 NEXT I1
42121 RETURN
```

Program IV.15: *Matrix save subroutine (MATSAVCB). Matrix C is saved in matrix B.*

```
42123 REM MATRIX SAVE (B IN C) SUBROUTINE (MATSAVBC)
42124 REM N1,N2 AND N3 ARE INPUT INDICES
42125 IF N1*N2*N3=0 THEN GOTO 42135
42126 REM CHECK DIMENSION
42127 FOR I1=1 TO N1
42128 FOR I2=1 TO N2
42129 FOR I3=1 TO N3
42130 C(I1,I2,I3)=B(I1,I2,I3)
42131 NEXT I3
42132 NEXT I2
42133 NEXT I1
42134 RETURN
42135 IF N1*N2=0 THEN GOTO 42142
42136 FOR I1=1 TO N1
42137 FOR I2=1 TO N2
42138 C(I1,I2)=B(I1,I2)
42139 NEXT I2
42140 NEXT I1
42141 RETURN
42142 IF N1=0 THEN RETURN
42143 FOR I1=1 TO N1
42144 C(I1)=B(I1)
42145 NEXT I1
42146 RETURN
```

Program IV.16: *Matrix save subroutine (MATSAVBC). Matrix B is saved in matrix C.*

```
42148 REM MATRIX SAVE (A IN C) SUBROUTINE (MATSAVAC)
42149 REM N1,N2 AND N3 ARE INPUT INDICES
42150 IF N1*N2*N3=0 THEN GOTO 42160
42151 REM CHECK DIMENSION
42152 FOR I1=1 TO N1
42153 FOR I2=1 TO N2
42154 FOR I3=1 TO N3
42155 C(I1,I2,I3)=A(I1,I2,I3)
42156 NEXT I3
42157 NEXT I2
42158 NEXT I1
42159 RETURN
42160 IF N1*N2=0 THEN GOTO 42167
42161 FOR I1=1 TO N1
42162 FOR I2=1 TO N2
42163 C(I1,I2)=A(I1,I2)
42164 NEXT I2
42165 NEXT I1
42166 RETURN
42167 IF N1=0 THEN RETURN
42168 FOR I1=1 TO N1
42169 C(I1)=A(I1)
42170 NEXT I1
42171 RETURN
```

Program IV.17: *Matrix save subroutine (MATSAVAC). Matrix A is saved in matrix C.*

```
42173 REM MATRIX SAVE (C IN A) SUBROUTINE (MATSAVCA)
42174 REM N1,N2 AND N3 ARE INPUT INDICES
42175 IF N1*N2*N3=0 THEN GOTO 42185
42176 REM CHECK DIMENSION
42177 FOR I1=1 TO N1
42178 FOR I2=1 TO N2
42179 FOR I3=1 TO N3
42180 A(I1,I2,I3)=C(I1,I2,I3)
42181 NEXT I3
42182 NEXT I2
42183 NEXT I1
42184 RETURN
42185 IF N1*N2=0 THEN GOTO 42192
42186 FOR I1=1 TO N1
42187 FOR I2=1 TO N2
42188 A(I1,I2)=C(I1,I2)
42189 NEXT I2
42190 NEXT I1
42191 RETURN
42192 IF N1=0 THEN RETURN
42193 FOR I1=1 TO N1
42194 A(I1)=C(I1)
42195 NEXT I1
42196 RETURN
```

Program IV.18: *Matrix save subroutine (MATSAVCA). Matrix C is saved in matrix A.*

Program IV.19:
```
REM PROGRAM TO DEMONSTRATE MATRIX OPERATIONS
PRINT"INPUT THE ROW AND COLUMN DIMENSIONS OF A(I,J) AND B(I,J):"
PRINT"ROW SIZE= ",
INPUT M
PRINT"COLUMN SIZE= ",
INPUT N
REM FIND MAXIMUM MATRIX DIMENSION NEEDED
IF M>N THEN K=M+1
IF N>=M THEN K=N+1
DIM A(K,K), B(K,K), C(K,K)
REM INPUT MATRIX A
PRINT
PRINT"INPUT MATRIX A ROW BY ROW"
PRINT
FOR I=1 TO M
PRINT"INPUT THE ",N," ELEMENTS OF ROW ",I," :"
FOR J=1 TO N
INPUT A(I,J)
NEXT J
PRINT
NEXT I
PRINT
REM PRINT MATRIX A
REM MOVE A TO C
```

Program IV.19: (Cont.)

```
N1=M
N2=N
N3=0
GOSUB 42150
PRINT"MATRIX A="
PRINT
REM PRINTING SUBROUTINE
GOSUB 1020
PRINT
REM INPUT MATRIX B
PRINT
PRINT"INPUT MATRIX B ROW BY ROW"
PRINT
FOR I=1 TO M
PRINT"INPUT THE ",N," ELEMENTS OF ROW ",I," :"
FOR J=1 TO N
INPUT B(I,J)
NEXT J
PRINT
NEXT I
PRINT
REM MOVE B TO C
GOSUB 42125
PRINT
PRINT"MATRIX B="
PRINT
REM PRINTING SUBROUTINE
GOSUB 1020
PRINT
REM MATRIX ADDITION
GOSUB 41800
PRINT"A+B="
REM PRINTING SUBROUTINE
GOSUB 1020
REM MATRIX SUBTRACTION
GOSUB 41850
PRINT"A-B="
REM PRINTING SUBROUTINE
GOSUB 1020
REM MATRIX TRANSPOSE (A)
REM FIRST, SAVE B IN C
GOSUB 42125
REM PUT TRANSPOSE OF A IN B
GOSUB 41950
REM MOVE TRANSPOSE OF A TO A
N1=N
N2=M
GOSUB 42075
REM RETURN C TO B
N1=M
N2=N
GOSUB 42100
PRINT
PRINT"A(TRANSPOSE)="
```

```
REM SWITCH ROW AND COLUMN SIZES FOR THE TRANSPOSE PRINT
N2=M
N1=N
M=N1
N=N2
REM MOVE A TO C FOR PRINTING
GOSUB 42150
PRINT
REM PRINTING SUBROUTINE
GOSUB 1020
REM DETERMINE PRODUCT OF A(TRANSPOSE) AND B
M1=M
N1=N
M2=N
N2=M
REM MATRIX MULTIPLICATION SUBROUTINE
GOSUB 41900
PRINT"A(TRANSPOSE) X B="
PRINT
REM PRINTING SUBROUTINE
N=M
GOSUB 1020
PRINT
END
```

Program IV.19: *Program to demonstrate the matrix operations of addition, subtraction, transposition, and multiplication (programs IV.8 thru IV.18). The printing subroutine shown in program IV.20 is also used.*

```
1000 REM *********************
1010 REM PRINTING SUBROUTINE
1020 PRINT
1030 FOR I=1 TO M
1035 FOR J=1 TO N
1040 PRINT TAB(6*J),INT(C(I,J)*100)/100,
1050 NEXT J
1055 PRINT
1060 NEXT I
1070 PRINT
1080 RETURN
```

Program IV.20: *Matrix printing subroutine used in this chapter. It is not considered a library subroutine since it is somewhat specific and limited.*

```
INPUT THE ROW AND COLUMN DIMENSIONS OF A(I,J) AND B(I,J):
ROW SIZE= ?3
COLUMN SIZE= ?2

INPUT MATRIX A ROW BY ROW

INPUT THE  2 ELEMENTS OF ROW  1 :
?1
?1

INPUT THE  2 ELEMENTS OF ROW  2 :
?2
?2

INPUT THE  2 ELEMENTS OF ROW  3 :
?3
?3

MATRIX A=

        1    1
        2    2
        3    3

INPUT MATRIX B ROW BY ROW

INPUT THE  2 ELEMENTS OF ROW  1 :
?4
?4

INPUT THE  2 ELEMENTS OF ROW  2 :
?5
?5

INPUT THE  2 ELEMENTS OF ROW  3 :
?6
?6

MATRIX B=

        4    4
        5    5
        6    6

A+B=

        5    5
        7    7
        9    9

A-B=

       -3   -3
       -3   -3
       -3   -3

A(TRANSPOSE)=

        1    2    3
        1    2    3

A(TRANSPOSE) X B=

       32   32
       32   32
```

Listing IV.2a: *Examples of matrix operations using program IV.19.*

Listing IV.2b:
```
INPUT THE ROW AND COLUMN DIMENSIONS OF A(I,J) AND B(I,J):
ROW SIZE= ?4
COLUMN SIZE= ?4

INPUT MATRIX A ROW BY ROW

INPUT THE   4 ELEMENTS OF ROW  1 :
?1
?0
?0
?0

INPUT THE   4 ELEMENTS OF ROW  2 :
?0
?1
?0
?0

INPUT THE   4 ELEMENTS OF ROW  3 :
?0
?0
?1
?0

INPUT THE   4 ELEMENTS OF ROW  4 :
?0
?0
?0
?1




MATRIX A=

         1        0        0        0
         0        1        0        0
         0        0        1        0
         0        0        0        1


INPUT MATRIX B ROW BY ROW

INPUT THE   4 ELEMENTS OF ROW  1 :
?1
?2
?3
?4

INPUT THE   4 ELEMENTS OF ROW  2 :
?5
?6
?7
?8

INPUT THE   4 ELEMENTS OF ROW  3 :
?9
?10
?11
?12

INPUT THE   4 ELEMENTS OF ROW  4 :
?13
?14
?15
?16
```

MATRIX B=

```
 1    2    3    4
 5    6    7    8
 9   10   11   12
13   14   15   16
```

A+B=

```
 2    2    3    4
 5    7    7    8
 9   10   12   12
13   14   15   17
```

A−B=

```
  0   -2   -3   -4
 -5   -5   -7   -8
 -9  -10  -10  -12
-13  -14  -15  -15
```

A(TRANSPOSE)=

```
1  0  0  0
0  1  0  0
0  0  1  0
0  0  0  1
```

A(TRANSPOSE) X B=

```
 1    2    3    4
 5    6    7    8
 9   10   11   12
13   14   15   16
```

Listing IV.2b: *Examples of matrix operations using program IV.19.*

IV.3 Other Matrix Operations

In this section we consider several more matrix subroutines which perform some of the standard functions discussed in textbooks on the subject. The programs presented can be used for mathematically oriented exercises involving detailed manipulation of matrix rows and columns, such as performing Gauss-Jordan elimination, triangularizing a matrix, or reorganizing a matrix for Gauss-Seidel iteration. For further information on the subject, see references 18, 22, 23, and 27.

One way to multiply every element in a matrix by a given constant is to create a diagonal matrix (a constant times the identity matrix; see subroutine 42000, MATSCALE), and premultiply by it. A faster way is to simply multiply each element in the matrix by the desired constant. The subroutine shown in program IV.21 does just that and is about five times quicker than the diagonal-matrix generation/multiplication approach.

This same subroutine could be used to clear a matrix by multiplying each element by zero. The multiplication step is unnecessary, and a faster subroutine which does not require this step is given in program IV.22.

Another important operation entails the switching of rows with one another and the corresponding switching of columns. Because the matrix-transpose subroutine switches rows with columns, we need only a program to interchange rows with one another. For example, to switch columns I and J, first transpose the matrix, switch rows I and J, and then transpose back. A subroutine for performing the row-switching function appears in program IV.23.

Another row and column operation entails the adding of a scalar multiple of one row to another row. The subroutine in program IV.24 adds the product of B and row N1 to row N2. These subroutines require specific information for proper operation. For further information, refer to the associated subroutine connection diagrams.

The subroutines given in this section are demonstrated by program IV.25, with a sample run shown in listing IV.3. Note that program IV.25 uses the same printing subroutine employed by program IV.19. Also observe that as a calling program, program IV.25 takes the responsibility for properly dimensioning the arrays.

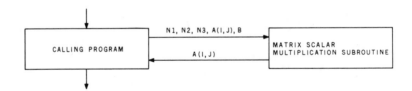

Figure IV.11. *Subroutine connection for multiplying a matrix A(I,J) by a scalar B (MATSCALE).*

Statements/Functions List

*

IF/THEN, FOR/NEXT

Variables List

A( ) (dimension determined by N1, N2, N3)
B, I1, I2, I3, N1, N2, N3

Variables Passed to Subroutine

A( ), B, N1, N2, N3

Table IV.13 *Statements, functions, and variables used in the matrix scalar-multiplication subroutine (MATSCALE).*

```
42198 REM SCALAR B X MATRIX A SUBROUTINE (MATSCALE)
42199 REM N1,N2 AND N3 ARE INPUT INDICES
42200 IF N1*N2*N3=0 THEN GOTO 42210
42201 REM CHECK DIMENSION
42202 FOR I1=1 TO N1
42203 FOR I2=1 TO N2
42204 FOR I3=1 TO N3
42205 A(I1,I2,I3)=B*A(I1,I2,I3)
42206 NEXT I3
42207 NEXT I2
42208 NEXT I1
42209 RETURN
42210 IF N1*N2=0 THEN GOTO 42217
42211 FOR I1=1 TO N1
42212 FOR I2=1 TO N2
42213 A(I1,I2)=B*A(I1,I2)
42214 NEXT I2
42215 NEXT I1
42216 RETURN
42217 IF N1=0 THEN RETURN
42218 FOR I1=1 TO N1
42219 A(I1)=B*A(I1)
42220 NEXT I1
42221 RETURN
```

Program IV.21: *Subroutine for multiplying a matrix by a scalar (MATSCALE).*

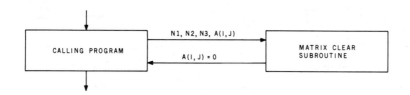

Figure IV.12. *Subroutine connection for clearing a matrix (MATCLRA).*

Statements/Functions List

*

IF/THEN, FOR/NEXT

Variables List

A( ), I1, I2, I3, N1, N2, N3

Variables Passed to Subroutine

A( ), N1, N2, N3

Table IV.14 *Statements, functions, and variables used in the matrix clearing subroutine (MATCLRA).*

```
42223 REM MATRIX A CLEAR SUBROUTINE (MATCLRA)
42224 REM N1,N2 AND N3 ARE INPUT INDICES
42225 IF N1*N2*N3=0 THEN GOTO 42235
42226 REM CHECK DIMENSION
42227 FOR I1=1 TO N1
42228 FOR I2=1 TO N2
42229 FOR I3=1 TO N3
42230 A(I1,I2,I3)=0
42231 NEXT I3
42232 NEXT I2
42233 NEXT I1
42234 RETURN
42235 IF N1*N2=0 THEN GOTO 42242
42236 FOR I1=1 TO N1
42237 FOR I2=1 TO N2
42238 A(I1,I2)=0
42239 NEXT I2
42240 NEXT I1
42241 RETURN
42242 IF N1=0 THEN RETURN
42243 FOR I1=1 TO N1
42244 A(I1)=0
42245 NEXT I1
42246 RETURN
```

Program IV.22: *Subroutine for clearing a matrix A(I,J) (MATCLRA).*

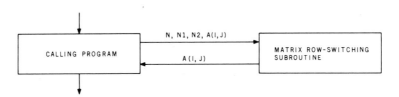

Figure IV.13. *Subroutine connection for switching rows in a matrix (MATSWCH).*

Statements/Functions List

FOR/NEXT

Variables List

A(N1,J), B, N, N1, N2

Variables Passed to Subroutine

A(I,J), N1, N2

Table IV.15 *Statements, functions, and variables used in the matrix row-switching subroutine (MATSWCH).*

```
42248 REM ROW SWITCHING SUBROUTINE (MATSWCH)
42249 REM ROWS N1 AND N2 ARE INTERCHANGED
42250 FOR J=1 TO N
42251 B=A(N1,J)
42252 A(N1,J)=A(N2,J)
42253 A(N2,J)=B
42254 NEXT J
42255 RETURN
```

Program IV.23: *Matrix row-switching subroutine (MATSWCH).*

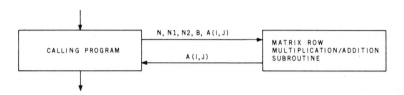

Figure IV.14. *Subroutine connection for adding B times row N1 to row N2 (MATRMAD).*

Statements/Functions List

+,*

FOR/NEXT

Variables List

A(I,J), B, J, N, N1, N2

Variables Passed to Subroutine

A(I,J), B, N, N1, N2

Table IV.16 *Statements, functions, and variables used in the matrix row-addition subroutine (MATRMAD).*

```
42273 REM ROW MULTIPLICATION/ADD SUBROUTINE (MATRMAD)
42274 REM B TIMES ROW N1 ADDED TO N2
42275 FOR J=1 TO N
42276 A(N2,J)=A(N2,J)+B*A(N1,J)
42277 NEXT J
42278 RETURN
```

Program IV.24: *Matrix subroutine to add the product of B times row N1 to row N2 (MATRMAD).*

Program IV.25:
```
REM PROGRAM TO DEMONSTRATE SPECIAL MATRIX OPERATIONS
REM FIRST CREATE A DIAGONAL MATRIX
PRINT"WHAT SIZE MATRIX IS DESIRED",
INPUT N
PRINT
PRINT"THIS MATRIX WILL HAVE ONLY DIAGONAL ELEMENTS.",
PRINT" WHAT VALUE SHOULD THEY BE",
INPUT B
DIM A(N+1,N+1),B(N+1,N+1),C(N+1,N+1)
REM GO TO DIAGONAL MATRIX CREATION SUBROUTINE
GOSUB 42000
REM SAVE MATRIX IN BOTH,A AND C
N1=N
N2=N
N3=0
GOSUB 42075
GOSUB 42125
REM PRINT RESULT
M=N
GOSUB 1020
REM SWITCH TWO ROWS
PRINT
PRINT"INPUT THE NUMBERS OF THE TWO ROWS TO BE SWITCHED",
INPUT N1,N2
REM GO TO SWITCHING SUBROUTINE
GOSUB 42250
REM MOVE RESULT TO C AND PRINT
N1=N
N2=N
GOSUB 42150
PRINT
GOSUB 1020
PRINT"INPUT SCALE FACTOR TO BE MULTIPLIED BY: ",
INPUT B
GOSUB 42200
REM PRINT RESULT
GOSUB 42150
GOSUB 1020
PRINT
PRINT"INPUT WHAT FRACTION OF ROW N1 IS TO BE ADDED TO ROW N2"
PRINT"FRACTION: ",
INPUT B
PRINT"N1: ",
INPUT N1
PRINT"N2: ",
INPUT N2
REM GO TO OPERATION
GOSUB 42275
REM PRINT RESULT
PRINT
N2=N
N1=N
GOSUB 42150
GOSUB 1020
```

```
PRINT"CLEAR MATRIX"
N1=N
N2=N
GOSUB 42225
REM PRINT RESULT
PRINT
GOSUB 42150
GOSUB 1020
END
```

Program IV.25: *Program to demonstrate creation of a diagonal matrix, switching two rows, adding two rows, and finally clearing the matrix (programs IV.12, IV.22, IV.23, and IV.24). The matrix save routines are used extensively, particularly to move results to matrix C(I,J) for printing.*

```
WHAT SIZE MATRIX IS DESIRED?5
THIS MATRIX WILL HAVE ONLY DIAGONAL ELEMENTS. WHAT VALUE SHOULD THEY BE?3

         3    0    0    0    0
         0    3    0    0    0
         0    0    3    0    0
         0    0    0    3    0
         0    0    0    0    3

INPUT THE NUMBERS OF THE TWO ROWS TO BE SWITCHED?1,4

         0    0    0    3    0
         0    3    0    0    0
         0    0    3    0    0
         3    0    0    0    0
         0    0    0    0    3

INPUT SCALE FACTOR TO BE MULTIPLIED BY: ?2.5

         0    0    0    7.5  0
         0    7.5  0    0    0
         0    0    7.5  0    0
         7.5  0    0    0    0
         0    0    0    0    7.5

INPUT WHAT FRACTION OF ROW N1 IS TO BE ADDED TO ROW N2
FRACTION: ?2
N1: ?3
N2: ?2

         0    0    0    7.5  0
         0    7.5  15   0    0
         0    0    7.5  0    0
         7.5  0    0    0    0
         0    0    0    0    7.5

CLEAR MATRIX

         0    0    0    0    0
         0    0    0    0    0
         0    0    0    0    0
         0    0    0    0    0
         0    0    0    0    0

READY
```

Listing IV.3: *Printout of sample results from program IV.25.*

IV.4 Matrix Coordinate Changes

The matrix subroutines may be used to perform coordinate changes in the form of scaling, rotating, and shifting. These coordinate transformations are useful in navigational and mechanical problems. For example, when dealing with the relative motions between a collection of moving particles (eg: charged masses), it is often very useful to perform the analysis in the center-of-mass frame of reference and subsequently transform the results back into the laboratory coordinate system. The following discussion shows how this can be done in principle.

Positions in the original coordinate system can be represented by the column vector:

$$V = \begin{bmatrix} x \\ y \\ z \end{bmatrix} \qquad \text{(IV.15)}$$

This rectangular coordinate system can be scaled to

$$V' = \begin{bmatrix} ax \\ by \\ cz \end{bmatrix} \qquad \text{(IV.16)}$$

by using the matrix S:

$$S = \begin{bmatrix} a & 0 & 0 \\ 0 & b & 0 \\ 0 & 0 & c \end{bmatrix} \qquad \text{(IV.17)}$$

This would be performed by the matrix multiplication subroutine in the symbolic form, $V' = S \times V$, by a subroutine call. Note, as used here, "$\times$" refers to the matrix product.

Rotating the coordinate system can also be performed by matrix multiplication. For a rotation of u radians about the Z axis the rotation matrix is:

$$U(u) = \begin{bmatrix} \cos(u) & \sin(u) & 0 \\ -\sin(u) & \cos(u) & 0 \\ 0 & 0 & 1 \end{bmatrix} \qquad \text{(IV.18)}$$

Observe that if the rotation angle is $u = 0$, then the rotation matrix becomes I, the identity matrix, and the coordinate system is left unchanged.

A similar matrix for rotation about the Y axis is:

$$V(v) = \begin{bmatrix} \cos(v) & 0 & \sin(v) \\ 0 & 1 & 0 \\ -\sin(v) & 0 & \cos(v) \end{bmatrix} \quad \text{(IV.19)}$$

For rotation about the X axis:

$$W(w) = \begin{bmatrix} 1 & 0 & 0 \\ 0 & \cos(w) & \sin(w) \\ 0 & -\sin(w) & \cos(w) \end{bmatrix} \quad \text{(IV.20)}$$

A coordinate system can be rotated to any angular position by applying the three rotation matrices shown above. Note, however, that only two rotations are required to do the job, not three. This can be understood by considering that only two angles were involved in the spherical coordinate system discussed earlier. All rotations can be performed relative to those two angles, which in our case are u and w.

The final coordinate change operation is that of shifting. This is simply the addition of a constant vector **P**.

The above coordinate transformation operations can be combined to give a complete linear coordinate change of the following form:

$$\mathbf{V'} = S \times W(w) \times U(u) \times \mathbf{V} + \mathbf{P} \quad \text{(IV.21)}$$

In words, the vector **V** (unprimed coordinate system with the vector starting at the origin) is first rotated about the Z axis through an angle u, and then about the X axis through an angle w. The resulting rotated coordinate system is then scaled by the diagonal matrix S, and finally the constant vector **P** is added to the result. Note that the order in which these operations are performed is very important; these operations cannot be switched without influencing the final coordinate change.

The utility of the above dissection of the matrix coordinate change procedure is simple. Coordinate transformations often appear very complicated. However, they can be broken down into conceptually simple operations, as shown above. The matrix or vector for each operation can be easily constructed, and the subroutines presented earlier can be employed to generate the final transformation. This clearly extends the use of the matrix subroutine library.

IV.5 Determinants

In this section, subroutines are presented for evaluating the determinant of a square matrix. For a mathematical discussion of determinants, the reader is directed to reference 27 or any similar text on matrix algebra.

The determinant has theoretical importance in solving simultaneous linear equations. In matrix algebra theory the solution to a linear set of equations can be written in terms of the determinant of the matrix representing the coefficients in the equations. This means of solving such problems is often referred to as Kramer's Rule (reference 27). Since the determinant appears in the denominator of the solution, the only way this approach can fail is if the determinant is zero (ie: if the *rank* of the coefficient matrix is less than N, where N is the row/column size of the square matrix). If the determinant is not zero, a solution exists at least by Kramer's Rule. However, as we will see, the solution can be attained more efficiently by other means. Thus, the determinant is generally not used in solving the problem, but rather is employed to check to see if a solution exists before proceeding further.

This existence test has an important extension to the matrix algebra problem of determining eigenvalues, as will be discussed in a later section. For the present, only the subroutines needed to evaluate the determinant will be considered.

Computer programs for evaluating determinants can be found in references 12 (2 by 2 matrix) and 25 (3 by 3 matrix). In this section, a general approach to determinant evaluation is considered, and a subroutine which handles matrices ranging from 1 by 1 to 4 by 4 is given.

The determinant corresponding to any square matrix can be found through the use of *cofactors*. A cofactor is a matrix derived from the original matrix by removing any one row and any one column. Thus, COFACTOR(A(I,J)) is an (N − 1) by (N − 1) matrix derived from matrix A, which is N by N, by removing row I and column J. In the subroutines presented in this book, only the cofactors corresponding to the first row of matrix A will be considered. This leaves only N cofactors to be treated, one for each column. In order to reserve I and J for other index use, K will be used to denote the particular row 1 cofactor. Thus, the Kth cofactor matrix is that which is derived from matrix A by removing the first row and the Kth column. The subroutine appearing in program IV.26 accomplishes this. See also figure IV.15.

The matrix cofactor is used in evaluating the determinant in the following manner:

$$\text{DET}(A) = \sum_{K=1}^{N} A(1,K) \times \text{SIGN} \times \text{DET}(\text{COFACTOR}(K)) \qquad (IV.22)$$

In the above equation A(1,K) is the Kth element of the first row of matrix A; SIGN is +1 or −1, depending on whether K, the column index, is odd or even, respectively; and COFACTOR(K) is the matrix cofactor belonging to the first row and Kth column of matrix A. This equation is true regardless of the dimension of A, thus providing a general algorithm for evaluating determinants.

A determinant-evaluation subroutine is presented in program IV.27 for calculating DET(A) for N = 1 through 4. For matrices up to 3 by 3, the evaluation is performed using direct algebraic expansions. For N = 4 the matrix-cofactor subroutine is called (see figure IV.16). Higher-dimensioned matrices can be evaluated by additional nested subroutine calls.

The determinant evaluation subroutine is demonstrated in program IV.28, with sample results shown in listing IV.4. Observe in listing IV.4b that the determinant of a diagonal matrix is simply the product of the diagonal elements.

As cautioned in the previous sections, care must be taken to ensure that the appropriate parameters are passed to the subroutine. Also, consult tables IV.17 and IV.18 to make certain that the variables employed by the subroutines are not in conflict with the calling program.

An additional concern in dealing with matrix determinants and inverting matrices (see next section) is accuracy. It is possible to encounter *ill-conditioned* matrices which have the annoying property of containing large coefficients (and perhaps some sign alternation), such that the evaluation involves taking the differences between large numbers. This can lead to large round-off errors in the final result. Two general approaches exist for circumventing this problem: preconditioning the matrix, which is often complicated, or using double precision. The latter approach is easiest, but not guaranteed.

As mentioned earlier, the determinant can be used in Kramer's Rule to solve simultaneous linear equations. However, this method is generally not employed. The more efficient matrix inversion approach discussed in the next two sections is preferable.

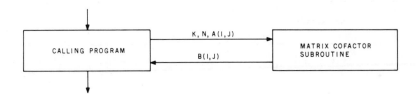

Figure IV.15. *Subroutine connection for determining the Kth cofactor of a square matrix (MATCOFAT).*

Statements/Functions List

+, −

IF/THEN, FOR/NEXT

Variables List

A(I,J), B(I,J), I, J, K, N

Variables Passed to Subroutine

A(I,J), K, N

Table IV.17 *Statements, functions, and variables used in the matrix cofactor subroutine (MATCOFAT).*

90 BASIC SCIENTIFIC SUBROUTINES

```
42296 REM COFACTOR K SUBROUTINE (MATCOFAT)
42297 REM INPUT MATRIX SIZE IS N X N
42298 REM MATRIX A(I,J) IN, MATRIX B(I,J) OUT
42299 REM FIRST SHIFT UP ONE ROW
42300 FOR I=2 TO N
42301 FOR J=1 TO N
42302 B(I-1,J)=A(I,J)
42303 NEXT J
42304 NEXT I
42305 FOR I=1 TO N-1
42306 FOR J=K TO N
42307 IF K=N THEN GOTO 42309
42308 B(I,J)=B(I,J+1)
42309 NEXT J
42310 NEXT I
42311 RETURN
```

Program IV.26: *Subroutine for determining the Kth cofactor of a square matrix A(I,J) (MATCOFAT).*

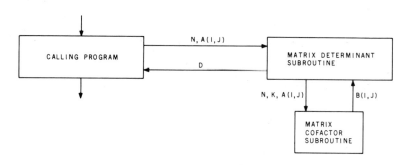

Figure IV.16. *Subroutine connection for matrix determinant evaluation (MATDET).*

Statements/Functions List

$+, -, *, >$
IF/THEN, FOR/NEXT, GOSUB

Variables List

A(I,J), C(I,J), D, D1, K, N, N1, N2, N3

Variables Passed to Subroutine

A(I,J), N

Table IV.18 *Statements, functions, and variables used in the determinant subroutine (MATDET).*

```
42348 REM MATRIX DETERMINANT SUBROUTINE (MATDET)
42349 REM FINDS DETERMINANT FOR UP TO A 4 X 4 MATRIX
42350 IF N>=2 THEN GOTO 42355
42351 REM ********************
42352 REM FIRST ORDER DETERMINANT
42353 D=A(1,1)
42354 RETURN
42355 IF N>=3 THEN GOTO 42360
42356 REM ********************
42357 REM SECOND ORDER DETERMINANT
42358 D=A(1,1)*A(2,2)-A(1,2)*A(2,1)
42359 RETURN
42360 IF N>=4 THEN GOTO 42370
42361 REM ********************
42362 REM THIRD ORDER DETERMINANT
42363 D=A(1,1)*(A(2,2)*A(3,3)-A(2,3)*A(3,2))
42364 D=D-A(1,2)*(A(2,1)*A(3,3)-A(2,3)*A(3,1))
42365 D=D+A(1,3)*(A(2,1)*A(3,2)-A(2,2)*A(3,1))
42366 RETURN
42367 REM ********************
42368 REM FOURTH ORDER DETERMINANT
42369 REM SAVE A IN C
42370 N1=N
42371 N2=N
42372 N3=0
42373 GOSUB 42150
42374 IF N>=5 THEN RETURN
42375 REM D1 WILL BE THE DETERMINANT
42376 D1=0
42377 REM FIND DETERMINANT OF EACH COFACTOR
42378 FOR K=1 TO 4
42379 REM GET COFACTOR K
42380 GOSUB 42300
42381 REM COFACTOR RETURNED IN B
42382 REM MOVE B TO A
42383 GOSUB 42075
42384 REM GET DET(A)
42385 GOSUB 42363
42386 D1=D1+C(1,K)*D
42387 REM REVERSE SIGN FOR NEXT COFACTOR
42388 D1=-D1
42389 REM SAVE C IN A
42390 GOSUB 42175
42391 NEXT K
42392 D=D1
42393 RETURN
```

Program IV.27: *Subroutine for calculating the determinant of matrices up to the fourth rank (MATDET).*

```
REM PROGRAM TO DEMONSTRATE DETERMINANT
PRINT"INPUT MATRIX SIZE: ",
INPUT N
DIM A(N+1,N+1),B(N+1,N+1),C(N+1,N+1)
PRINT
REM INPUT MATRIX
FOR I=1 TO N
PRINT"INPUT ROW ",I
FOR J=1 TO N
INPUT A(I,J)
NEXT J
PRINT
NEXT I
PRINT
PRINT "MATRIX A="
REM MOVE A TO C FOR PRINTING
N1=N
N2=N
N3=0
GOSUB 42150
REM PRINT
M=N
GOSUB 1020
PRINT
REM EVALUATE DETERMINANT
GOSUB 42500
PRINT"DETERMINANT= ",D
END
```

Program IV.28: *Program to demonstrate the operation of the matrix determinant subroutine (shown in program IV.27).*

```
INPUT MATRIX SIZE: ?2

INPUT ROW   1
?1
?8

INPUT ROW   2
?2
?5


MATRIX A=

        1     8
        2     5


DETERMINANT=   -11
```

Listing IV.4a: *Determinant example for a second rank matrix.*

```
INPUT MATRIX SIZE: ?4

INPUT ROW   1
?1
?0
?0
?0

INPUT ROW   2
?0
?2
?0
?0

INPUT ROW   3
?0
?0
?3
?0

INPUT ROW   4
?0
?0
?0
?4

MATRIX A=

        1       0       0       0
        0       2       0       0
        0       0       3       0
        0       0       0       4

DETERMINANT=    24
```

Listing IV.4b: *Determinant example for a diagonal matrix, fourth rank.*

```
INPUT MATRIX SIZE: ?4

INPUT ROW   1
?1
?2
?3
?4

INPUT ROW   2
?5
?6
?7
?8

INPUT ROW   3
?3
?2
?5
?4

INPUT ROW   4
?8
?7
?0
?1
```

Listing IV.4c:

```
MATRIX A=
    1    2    3    4
    5    6    7    8
    3    2    5    4
    8    7    0    1

DETERMINANT=   -32
```

Listing IV.4c: *Determinant example for an arbitrary fourth rank matrix.*

IV.6 Matrix Inversion

Matrix inversion is one of the most important operations existing in matrix algebra. It is either explicitly or implicitly utilized in many of the common methods for determining the solution to sets of simultaneous linear equations. In this section, a general subroutine is presented for matrix inversion. This subroutine is then used in the next section for solving linear equations.

A square matrix A has an inverse if and only if its determinant is not zero. The inverse of a matrix is such that when it either premultiplies or postmultiplies the original array, the identity matrix results. That is,

$$[\text{inverse}(A)] \times A = I \qquad \text{(IV.23)}$$
$$A \times [\text{inverse}(A)] = I$$

It is tempting to write inverse(A) = 1/A, but matrix division has no defined meaning. However, the analogy, if kept in perspective, is useful. For example, using this idea we can rationalize why the determinant must be non-zero for the inverse to exist.

For matrix dimensions $N = 1$ through $N = 3$, the inverse can be written in reasonably compact algebraic terms. A program for inverting a 2 by 2 matrix is given in reference 12. Programs for 3 by 3 matrices are shown in references 19 and 25. Matrices up to dimension 10 by 10 (and higher with a change) can be inverted by the program given in reference 17. The subroutine appearing in program IV.29 is similar to that in reference 17 and is based on Gauss-Jordan Elimination. A discussion of this method, although it is simple, is beyond the scope of this book and the reader is directed to reference 22 for an excellent analysis. An equivalent FORTRAN program is given in reference 18.

The subroutine connection diagram is given in figure IV.17. The subroutine can be exercised by program IV.30, with sample results shown in listing IV.5.

Listing IV.5a shows the inversion of a diagonal matrix. The inverse matrix is also diagonal, and the elements are simply the reciprocals of the corresponding elements in the original matrix. As a check, the product of the matrix and its inverse gives the identity matrix. The second matrix inverted (listing IV.5b) is triangular. Although the inverse looks curious, it is correct. The third example is arbitrary. The fourth matrix cannot be inverted because its determinant is zero. The failure observed in the last example is not the fault of the subroutine. Simply, the given problem has no solution.

As usual, for proper operation the appropriate parameters must be passed to the subroutine, and the matrices used must be previously dimensioned by the calling program.

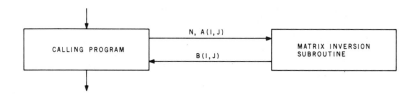

Figure IV.17. *Subroutine connection for matrix inversion (MATINV).*

Statements/Functions List

+, −, *, /, >
ABS, IF/THEN, FOR/NEXT

Variables List

A(I,J), B, B(I,J), I, J, K, M, N

Variables Passed to Subroutine

A(I,J), N

Table IV.19 *Statements, functions, and variables used in the matrix inversion subroutine (MATINV).*

```
42395 REM MATRIX INVERSION SUBROUTINE (MATINV)
42396 REM GAUSS-JORDAN ELIMINATION
42397 REM MATRIX A IS INPUT, MATRIX B IS OUTPUT
42398 REM DIM A=N X N    TEMPORARY DIM B=N X 2N
42399 REM FIRST CREATE MATRIX WITH A ON THE LEFT AND I ON THE RIGHT
42400 FOR I=1 TO N
42401 FOR J=1 TO N
42402 B(I,J+N)=0
```

Program IV.29:

```
42403 B(I,J)=A(I,J)
42404 NEXT J
42405 B(I,I+N)=1
42406 NEXT I
42407 REM PERFORM ROW ORIENTED OPERATIONS TO CONVERT THE LEFT HAND
42408 REM SIDE OF B TO THE IDENTITY MATRIX. THE INVERSE OF A WILL
42409 REM THEN BE ON THE RIGHT.
42410 FOR K=1 TO N
42411 IF K=N THEN GOTO 42424
42412 M=K
42413 REM FIND MAXIMUM ELEMENT
42414 FOR I=K+1 TO N
42415 IF ABS(B(I,K))>ABS(B(M,K)) THEN M=I
42416 NEXT I
42417 IF M=K THEN GOTO 42424
42418 FOR J=K TO 2*N
42419 B=B(K,J)
42420 B(K,J)=B(M,J)
42421 B(M,J)=B
42422 NEXT J
42423 REM DIVIDE ROW K
42424 FOR J=K+1 TO 2*N
42425 B(K,J)=B(K,J)/B(K,K)
42426 NEXT J
42427 IF K=1 THEN GOTO 42434
42428 FOR I=1 TO K-1
42429 FOR J=K+1 TO 2*N
42430 B(I,J)=B(I,J)-B(I,K)*B(K,J)
42431 NEXT J
42432 NEXT I
42433 IF K=N THEN GOTO 42441
42434 FOR I=K+1 TO N
42435 FOR J=K+1 TO 2*N
42436 B(I,J)=B(I,J)-B(I,K)*B(K,J)
42437 NEXT J
42438 NEXT I
42439 NEXT K
42440 REM RETRIEVE INVERSE FROM THE RIGHT SIDE OF B
42441 FOR I=1 TO N
42442 FOR J=1 TO N
42443 B(I,J)=B(I,J+N)
42444 NEXT J
42445 NEXT I
42446 RETURN
```

Program IV.29: *Matrix inversion subroutine (MATINV).*

```
REM PROGRAM TO DEMONSTRATE MATRIX INVERSION
PRINT"INPUT DIMENSION OF MATRIX TO BE INVERTED: ",
INPUT N
DIM A(N+1,N+1),B(N+1,2*N+2),C(N+1,N+1)
```

```
PRINT
FOR I=1 TO N
PRINT"INPUT ROW ",I," :"
FOR J=1 TO N
INPUT A(I,J)
NEXT J
PRINT
NEXT I
PRINT
PRINT"MATRIX A="
N1=N
N2=N
N3=0
REM MOVE A TO C
GOSUB 42150
M=N
GOSUB 1020
REM GOTO INVERSION SUBROUTINE
GOSUB 42400
REM MOVE B TO C TO PRINT
PRINT"INVERSE OF A="
PRINT
GOSUB 42125
M=N
GOSUB 1020
REM CHECK RESULTS
REM MULTIPLY A TIMES B
M1=N
M2=N
GOSUB 41900
M=N
REM PRINT RESULTS
PRINT
PRINT"MATRIX A TIMES INVERSE MATRIX A="
GOSUB 1020
END
```

Program IV.30: *Program to demonstrate matrix inversion using program IV.29.*

```
INPUT DIMENSION OF MATRIX TO BE INVERTED: ?3

INPUT ROW  1 :
?1
?0
?0

INPUT ROW  2 :
?0
?2
?0

INPUT ROW  3 :
?0
?0
?3
```

Listing IV.5a:

```
MATRIX A=

        1      0      0
        0      2      0
        0      0      3

INVERSE OF A=

        1      0      0
        0     .5      0
        0      0    .33

MATRIX A TIMES INVERSE MATRIX A=

        1      0      0
        0      1      0
        0      0      1
```

Listing IV.5a: *A sample run of program IV.30 to calculate the inverse of a diagonal matrix.*

Listing IV.5b:
```
INPUT DIMENSION OF MATRIX TO BE INVERTED: ?5
INPUT ROW  1 :
?1
?0
?0
?0
?0

INPUT ROW  2 :
?1
?1
?0
?0
?0

INPUT ROW  3 :
?1
?1
?1
?0
?0

INPUT ROW  4 :
?1
?1
?1
?1
?0

INPUT ROW  5 :
?1
?1
?1
?1
?1

MATRIX A=
        1      0      0      0      0
        1      1      0      0      0
        1      1      1      0      0
        1      1      1      1      0
        1      1      1      1      1
```

```
INVERSE OF A=

         1     0     0     0     0
        -1     1     0     0     0
         0    -1     1     0     0
         0     0    -1     1     0
         0     0     0    -1     1

MATRIX A TIMES INVERSE MATRIX A=

         1     0     0     0     0
         0     1     0     0     0
         0     0     1     0     0
         0     0     0     1     0
         0     0     0     0     1
```

Listing IV.5b: *A run of program IV.30 calculating the inverse of a lower triangular matrix of the fifth rank.*

```
INPUT DIMENSION OF MATRIX TO BE INVERTED: ?5

INPUT ROW   1 :
?1
?3
?2
?6
?4

INPUT ROW   2 :
?3
?5
?4
?2
?6

INPUT ROW   3 :
?3
?8
?6
?7
?5

INPUT ROW   4 :
?2
?5
?3
?4
?4

INPUT ROW   5 :
?7
?6
?8
?7
?6

MATRIX A=

         1     3     2     6     4
         3     5     4     2     6
         3     8     6     7     5
         2     5     3     4     4
         7     6     8     7     6
```

Listing IV.5c:

```
INVERSE OF A=

   -.26   -.37   -.6    1.18    .26
   -.28   -.19   -.07    .75   -.06
    .14    .39    .67  -1.48   -.05
    .16   -.18   -.04    .09    .04
    .22    .31   -.07   -.25   -.06

MATRIX A TIMES INVERSE MATRIX A=

    1      0      0      0      0
    0      1      0      0      0
    0      0      1      0      0
    0      0      0      1      0
    0      0      0      0      1
```

Listing IV.5c: *A run of program IV.30 calculating the inverse of an arbitrary fifth rank matrix.*

```
RUN

INPUT DIMENSION OF MATRIX TO BE INVERTED: ?5

INPUT ROW   1 :
?0
?1
?2
?3
?4

INPUT ROW   2 :
?4
?5
?6
?7
?8

INPUT ROW   3 :
?1
?3
?5
?7
?9

INPUT ROW   4 :
?1
?4
?7
?0
?3

INPUT ROW   5 :
?1
?5
?9
?4
?8

MATRIX A=

    0      1      2      3      4
    4      5      6      7      8
    1      3      5      7      9
    1      4      7      0      3
    1      5      9      4      8

DIVIDE ZERO ERROR IN LINE 42425
READY
```

Listing IV.5d: *A run of program IV.30 where there is a failure in the inverse routine. The matrix rank was less than five and the determinant was zero.*

IV.7 Solving Linear Sets of Equations

Many situations involving sets of equations are amenable to solution using the matrix inversion subroutine discussed in the previous section. For example, we might desire to determine the intersection of two (non-parallel) lines lying in the (X,Y) plane. The coordinates of the intersection must satisfy the equations defining each line, and thus the coordinates of the intersection are the solution to the set of linear equations. In this example the two linear equations are:

$$aX + bY = c$$
$$dX + eY = f \quad \text{(IV.24a)}$$

These equations can be written in a more matrix-oriented form as:

$$A \times X = B \quad \text{(IV.24b)}$$

where

$$A = \begin{bmatrix} a & b \\ d & e \end{bmatrix} \quad X = \begin{bmatrix} x \\ y \end{bmatrix} \quad B = \begin{bmatrix} c \\ f \end{bmatrix}$$

This is the equivalent of:

$$A(1,1) \times X(1,1) + A(1,2) \times X(2,1) = B(1,1)$$
$$A(2,1) \times X(1,1) + A(2,2) \times X(2,1) = B(2,1) \quad \text{(IV.24c)}$$

In the latter representation, A(I,J) may be considered a 2 by 2 coefficient matrix, X(I,1) a 2 by 1 variable vector, and B(I,1) a 2 by 1 constant vector. Since A is a square matrix, if the lines are not parallel, then there is a solution. The determinant of A must therefore be non-zero and the inverse of matrix A must exist. The solution is then:

$$[\text{inverse}(A)] \times A \times X = X = [\text{inverse}(A)] \times B \quad \text{(IV.25)}$$

The solution to the problem is the constant vector **B**, multiplied by the inverse of the coefficient matrix.

The above discussion need not be limited to two lines in a plane. It could easily be extended to an N-dimensional problem requiring N constitutive independent linear equations for solution. The matrix formulation is the same.

Two terms used in the preceding paragraphs require some elaboration. The first is *linear*. This simply refers to the form of the equations; they cannot have nonlinear variables such as X^2 or $\ln(X)$. However, in some cases it is possible to linearize a set of equations by transforming the variables. For example, consider the intersection of the two curves:

$$aX + b\ln(Z) = c$$
$$dX + e\ln(Z) = f$$
(IV.26)

These equations could be converted to a linear form by the variable change: $Y = \ln(Z)$.

The second term is *independent*. This refers to the condition that none of the N equations can be constructed from linear combinations of the other $N - 1$ equations. If this condition is violated, then there will not be enough information for a solution and the determinant of the coefficient matrix will be zero. The inverse will not exist. Such a matrix is called *singular*.

Because solving simultaneous equations is so important, there is vast literature on the subject. Programs can be found in references 12 and 16 for the case $N = 2$; in reference 25 for $N = 2$ and $N = 3$; and in reference 16 for $N = 4$.

The routine given in program IV.31 demonstrates the use of the matrix inversion subroutine, as well as several others, to solve simultaneous equations of any order. It is presented as a calling program and not as an executive subroutine since general subroutines are being used to handle a specific problem. Operation of the program is demonstrated in listings IV.6a and b. In the first example, a solution was chosen and the equations were then made up to test the operation of the program. In the second case, an arbitrary set of five-dimensional equations was entered. The computer solution is correct.

As with many of the matrix operations, solving simultaneous linear equations by matrix inversion is best performed in the double-precision mode.

Program IV.31:
```
REM PROGRAM TO DEMONSTRATE USE OF INVERSE TO SOLVE
REM SIMULTANEOUS EQUATIONS
PRINT"INPUT SIZE OF EQUATION (NUMBER OF UNKNOWNS): ",
INPUT N
DIM A(N+1,N+1),B(N+1,2*N+2),C(N+1,N+1)
PRINT
FOR I=1 TO N
PRINT"INPUT ROW ",I," OF COEFFICIENT MATRIX"
FOR J=1 TO N
INPUT A(I,J)
NEXT J
PRINT
GOSUB 2000
PRINT"INPUT CONSTANT WHICH ROW EQUATION EQUALS: ",
INPUT B(I,1)
PRINT
NEXT I
PRINT
PRINT"MATRIX A="
GOSUB 2000
N2=N
N3=0
```

```
REM MOVE A TO C
GOSUB 42150
REM PRINT A
M=N
GOSUB 1020
GOSUB 2000
REM MOVE CONSTANT VECTOR TO C
GOSUB 42125
REM PRINT CONSTANT VECTOR
PRINT
PRINT"CONSTANT VECTOR ="
N=1
GOSUB 1020
N=M
REM OBTAIN INVERSE
GOSUB 42400
REM MOVE RESULT IN B TO A
GOSUB 42075
REM MOVE CONSTANT VECTOR IN C TO B
GOSUB 42100
REM MULTIPLY INVERSE TIMES CONSTANT VECTOR
M1=N
N1=N
M2=N
N2=1
GOSUB 41900
REM RESULT IS IN C. PRINT C
M=N
N=1
PRINT"SOLUTION VECTOR="
GOSUB 1020
END
```

Program IV.31: *Program to demonstrate the application of the matrix inversion subroutine (program IV.29) to solve simultaneous linear equations.*

```
INPUT SIZE OF EQUATION (NUMBER OF UNKNOWNS): ?4

INPUT ROW   1 OF COEFFICIENT MATRIX
?1
?2
?3
?4

INPUT CONSTANT WHICH ROW EQUATION EQUALS: ?10

INPUT ROW   2 OF COEFFICIENT MATRIX
?1
?2
?3
?0

INPUT CONSTANT WHICH ROW EQUATION EQUALS: ?6
```

Listing IV.6a:

```
INPUT ROW   3 OF COEFFICIENT MATRIX
?1
?2
?0
?0

INPUT CONSTANT WHICH ROW EQUATION EQUALS: ?3

INPUT ROW   4 OF COEFFICIENT MATRIX
?1
?0
?0
?0

INPUT CONSTANT WHICH ROW EQUATION EQUALS: ?1

MATRIX A=

    1     2     3     4
    1     2     3     0
    1     2     0     0
    1     0     0     0

CONSTANT VECTOR =

    10
     6
     3
     1

SOLUTION VECTOR=

     1
     1
     1
     1
```

Listing IV.6a: *A sample run of program IV.31 using fourth order simultaneous equations.*

Listing IV.6b:
```
INPUT SIZE OF EQUATION (NUMBER OF UNKNOWNS): ?5

INPUT ROW   1 OF COEFFICIENT MATRIX
?2
?3
?5
?3
?4

INPUT CONSTANT WHICH ROW EQUATION EQUALS: ?8

INPUT ROW   2 OF COEFFICIENT MATRIX
?3
?4
?8
?0
?6

INPUT CONSTANT WHICH ROW EQUATION EQUALS: ?5

INPUT ROW   3 OF COEFFICIENT MATRIX
?2
?5
?3
?9
?6
```

```
INPUT CONSTANT WHICH ROW EQUATION EQUALS: ?7

INPUT ROW   4 OF COEFFICIENT MATRIX
?2
?3
?3
?4
?1

INPUT CONSTANT WHICH ROW EQUATION EQUALS: ?0

INPUT ROW   5 OF COEFFICIENT MATRIX
?3
?4
?9
?9
?1

INPUT CONSTANT WHICH ROW EQUATION EQUALS: ?4

MATRIX A=

        2     3     5     3     4
        3     4     8     0     6
        2     5     3     9     6
        2     3     3     4     1
        3     4     9     9     1

CONSTANT VECTOR =

        8
        5
        7
        0
        4

SOLUTION VECTOR=

        23.39
        -19.13
        -2.85
        3.36
        5.69
```

Listing IV.6b: *A run of program IV.31 using fifth order simultaneous equations.*

IV.8 Characteristic Polynomials and Eigenvalues

This and the following section pertain to eigenvalues and eigenvectors. Often a frightening area for the beginning engineering student, this subject is usually taught from a theoretical perspective. In simple terms, eigenvalues and eigenvectors are particular solutions to the matrix problem:

$$A \times X = gX \qquad (IV.27)$$

where A is an N by N coefficient matrix (given), X is an N by 1 vector (to be determined), and g is a constant (also to be determined). Curiously, there are ways to calculate the eigenvalue g without determining the eigenvector **X** and vice versa.

Eigenvalues are often encountered in electrical and mechanical engineering problems involving oscillations. As an illustration, consider the problem of two masses suspended by springs. Let the first mass, M_1, be suspended from the ceiling (ie: a rigid support) by a spring having an elastic constant K_1. Let the second mass, M_2, be suspended from the first mass by a spring having an elastic constant K_2. Further, let the positions of the two masses relative to their undisturbed locations be X_1 and X_2, respectively. If the system is disturbed, the balance of force equations are:

$$M_1 d^2(X_1)/dt^2 = K_1 X_1 - K_2(X_2 - X_1)$$
$$M_2 d^2(X_2)/dt^2 = K_2(X_2 - X_1) \qquad \text{(IV.28)}$$

In words, the inertial force as determined from the acceleration of the masses (ie: the second order time derivative) must be balanced by the applied force (ie: the springs). This is Newton's law of motion. We expect the resulting motion to be oscillatory, and thus describable by functions of the form:

$$X_1 = \text{constant} \times \sin(wt)$$
$$X_2 = \text{constant} \times \sin(wt) \qquad \text{(IV.29)}$$

The constant contains the amplitude and phase information, and w is the frequency of oscillation in radians per second.

Since the second order time derivatives of X_1 and X_2 are just $-w^2 X_1$ and $-w^2 X_2$, respectively, the equations of motion may be written in the form:

$$A(1,1) \times X(1,1) + A(1,2) \times X(2,1) = w^2 X(1,1)$$
$$A(2,1) \times X(1,1) + A(2,2) \times X(2,1) = w^2 X(2,1) \qquad \text{(IV.30)}$$

The coefficients, $A(I,J)$, are functions of only K_1, K_2, M_1, and M_2. These equations have the same form as the eigenvalue equation considered earlier in this section, but with $w^2 = g$.

There are two eigenvalues which serve as solutions to this problem. Later in this section we will discuss how they can be obtained. Corresponding to these eigenvalues are two eigenvectors. The meaning of these eigenvectors can be explained as follows. For a particular eigenvector, the ratio of the amplitudes of the oscillations of the two masses is a constant, and they maintain this amplitude ratio as they oscillate at the same frequency. If the ratio of the amplitudes is 2:1 for one of the eigenvectors, denoted as **V1**, then one *mode* of oscillation is:

$$\mathbf{X} = \begin{bmatrix} X_1 \\ X_2 \end{bmatrix} = \begin{bmatrix} 1 \\ 2 \end{bmatrix} c \times \sin(w_1 t) = \mathbf{V1} \times \sin(w_1 t) \qquad \text{(IV.31)}$$

Note that w_1^2 is the eigenvalue corresponding to that eigenvector. As the vector **X** appears on both sides of the original eigenvalue equation, any multiple of the eigenvector is still an eigenvector. The convention is to normalize the vector obtained. This convention is followed in the next section. The conclusion is that each eigenvalue frequency corresponds to a *normal mode* of oscillation in which both masses oscillate in unison, but not necessarily in phase with one another with the same frequency. Because the system is linear, the amplitude can be scaled to any level.

For this mass and spring problem, two such normal modes exist. However, common sense indicates that arbitrarily disturbing such a system will not lead to such organized motion. Instead, compound motion, a superposition of motions corresponding to both frequencies, will occur. If **V1** is the normalized eigenvector belonging to w_1 and **V2** the normalized eigenvector belonging to w_2, then it can be shown that any legitimate motion of the system can be represented by

$$X = a \times V1 \times \sin(w_1 t) + b \times V2 \times \sin(w_2 t) \tag{IV.32}$$

where a and b are constants which satisfy conditions specified elsewhere (*boundary conditions*).

From the above example it can be seen how a physical system may be described as an eigenvalue problem, thus permitting the use of standardized solution techniques. Observe that the dimension of the coefficient matrix need not be 2; it can be much larger. Also note that the characteristics of the physical situation are all contained in that matrix, and, in general, all possible solutions can be described as sums of the eigenvectors multiplied by their sinusoidal time factors (ie: they form a *complete* set).

Up to this point we have discussed only the origin and meaning of the eigenvalue/eigenvector equation and its solutions. Obtaining those solutions is often quite a task. In the remainder of this section we will consider a somewhat uncommon but practical approach to obtaining eigenvalues—regression analysis.

With the help of the identity matrix, the eigenvalue equation can be rewritten as:

$$(A - gI) \times X = 0 \tag{IV.33}$$

If both sides of that equation are premultiplied by the inverse of $(A - gI)$, we would get $X = 0$ as the solution for all problems. That cannot be true. Thus, the inverse used must not exist, meaning:

$$\text{DET}(A - gI) = 0 \tag{IV.34}$$

We define the following function:

$$y(x) = \text{DET}(A - xI) \tag{IV.35}$$

The desired eigenvalues are the roots of $y(x)$. Thus, the problem is reduced to obtaining the function y(x) and then finding the roots. Determining y(x) can be done as follows.

For a three-dimensional problem, the matrix $A - xI$ is:

$$A - xI = \begin{bmatrix} A(1,1) - x & A(1,2) & A(1,3) \\ A(2,1) & A(2,2) - x & A(2,3) \\ A(3,1) & A(3,2) & A(3,3) - x \end{bmatrix} \quad (IV.36)$$

Taking the determinant of this matrix results in a third degree polynomial of the form:

$$y(x) = ax^3 + bx^2 + cx + d \quad (IV.37)$$

The four coefficients can be found by calculating the determinant for four different guessed values of x, setting up four simultaneous equations and solving as demonstrated in the previous section. Alternatively, the four sets of (x,y) pairs can be treated as data and curve-fitted with a third-degree polynomial; the regression fit would be perfect. The latter approach will be discussed.

The above determinant argument can be generalized to any size coefficient matrix, A. If the dimension (or *rank*) of the matrix is N, then the *characteristic* polynomial y(x) will also be of degree N and will have N roots (possibly complex). The polynomial may be obtained from a regression analysis of N + 1 guessed points.

Examples of this procedure appear in listings IV.7a and IV.7b. In both cases, the matrix determinant program (program IV.28) is repeatedly used with different values of x. One of the guesses is always x = 0.

In the first example, the N = 3 matrix is probed with x equal to 0, −1, −2, and −3. In two of the cases, a zero determinant is obtained, which immediately indicates that two of the three eigenvalues were hit; g = −1 and g = −2. Using the four (x,y) pairs obtained, the following regression equation can be generated:

$$y(x) = x^3 + 2x^2 - x - 2 \quad (IV.38)$$

The above regression equation was determined using a program given in Volume II. Also refer to the Nth order polynomial regression program given in reference 17, as well as discussions in references 1, 2, 8, 21, and 22. The above third-degree polynomial can be factored as shown below:

$$y(x) = (x + 1)(x + 2)(x - 1) \quad (IV.39)$$

The roots, and therefore eigenvalues, are 1, −1, and −2. If the polynomial had not been easy to factor, one of the zero-search subroutines given in Volume II could have been used.

The second example, listing IV.7b, deals with an arbitrary fourth-order matrix. Five (x,y) pairs are needed to generate the fourth-degree characteristic polynomial shown below:

$$y(x) = x^4 - 18x^3 - 573x^2 + 4057x - 1864 \qquad \text{IV.40}$$

From the above examples, it is apparent that any eigenvalue problem can be approached from a characteristic polynomial perspective using polynomial-regression analysis. The linear simultaneous-equation technique can also be used to generate the polynomial, but with a little more effort. However, the methods are restricted to matrices having real (not complex) coefficients, though the eigenvalues themselves may be complex.

There is art involved in choosing the x values. In principle, any set of x values can be used as probes as long as there are no repeats. However, the accuracy of the results derived from the regression is not perfect and can be poor if the chosen x values are close together and far from any of the roots of the true polynomial. Therefore, some spread is helpful, particularly with respect to bracketing the roots. Volume II examines various techniques for determining the roots of a function.

One approach that can be used to greatly improve the accuracy of the calculation is to iterate. With this method, the above procedure is repeated using the eigenvalues calculating as new guesses for the x values, resulting in even better eigenvalue estimates. Iteration can be continued until the computer round-off error limit is reached. This approach is considered in detail in Volume II.

Listing IV.7a:

```
MATRIX A=

        1     0     0
        0    -1     0
        0     0    -2

DETERMINANT=   2



MATRIX A=

        2     0     0
        0     0     0
        0     0    -1

DETERMINANT=   0



MATRIX A=

        3     0     0
        0     1     0
        0     0     0
```

DETERMINANT= 0

MATRIX A=

    4   0   0
    0   2   0
    0   0   1

DETERMINANT= 8

Listing IV.7a: *Use of the determinant subroutine (see program IV.27) to probe for the characteristic polynomial. The eigenvalue guesses tried were 0, −1, −2, and −3.*

Listing IV.7b:

MATRIX A=

    1   3   5   2
    2   3   7   6
    2   89   6   5
    3   5   1   8

DETERMINANT= -1864

MATRIX A=

    0   3   5   2
    2   2   7   6
    2   89   5   5
    3   5   1   7

DETERMINANT= 1603

MATRIX A=

    2   3   5   2
    2   4   7   6
    2   89   7   5
    3   5   1   9

DETERMINANT= -6475

MATRIX A=

   -1   3   5   2
    2   1   7   6
    2   89   4   5
    3   5   1   6

DETERMINANT= 3830

```
MATRIX A=
    -2    3    5    2
     2    0    7    6
     2   89    3    5
     3    5    1    5

DETERMINANT=   4745
```

Listing IV.7b: *Use of the determinant subroutine (see program IV.27) to probe for the characteristic polynomial. The eigenvalue guesses tried were 0, 1, −1, 2, and 3.*

IV.9 Eigenvalues by the Power Method

Eigenvalues and eigenvectors were introduced in the previous section using the example of a two-spring oscillating system. It was shown that the natural frequencies of oscillation were directly related to the eigenvalues, which in turn were determined by the coefficient matrix. A regression technique was also discussed for generating the characteristic polynomial that can be used to calculate the eigenvalues.

In many situations, only limited information is needed regarding the frequencies of vibration. For example, in an automobile cabin design, the desire might be to find the highest frequency of natural vibration (largest eigenvalue) to see if there would be annoying, audible drumming caused by driving over rough roads. In other situations, the lowest frequency (smallest eigenvalue) may be of interest.

The largest or smallest eigenvalue belonging to a given coefficient matrix can be determined using one of the properties of eigenvectors: completeness. Following the earlier spring and mass analogy, any unforced vibration of the system can be represented by a vector **Y**, which can be described by the eigenvectors of the system:

$$Y = a_1 X_1 + a_2 X_2 + a_3 X_3 + \cdots + a_N X_N \qquad (IV.41)$$

In this representation, X_i is the ith eigenvector which obeys the eigenequation $A \times X_i = L_i X_i$, where L_i is the corresponding eigenvalue. Multiplying the general vector **Y** by the coefficient matrix A gives:

$$A \times Y = a_1 L_1 X_1 + a_2 L_2 X_2 + a_3 L_3 X_3 + \cdots + a_N L_N X_N \qquad (IV.42)$$

Multiplying again:

$$A \times A \times Y = a_1 L_1 L_1 X_1 + a_2 L_2 L_2 X_2 + a_3 L_3 L_3 X_3 + \cdots + a_N L_N L_N X_N \qquad (IV.43)$$

There is almost always a largest single eigenvalue, and therefore one term will dominate as the multiplication is repeated. This dominant term will correspond to the largest eigenvalue.

For further discussion, see references 7 and 27. FORTRAN programs using the repeated multiplication approach are given in references 18 (for real symmetric matrices) and 22. A BASIC subroutine employing the concept is shown in program IV.32.

Operation of this subroutine is fairly simple. The calling program passes to the routine the coefficient matrix, A, the size of the matrix, N, the accuracy desired, E, and the maximum number of iterations allowed, D1. The latter parameter ensures that the subroutine will not hang up if rapid convergence does not occur.

The subroutine returns to the calling program an estimate of the eigenvalue, the number of iterations actually performed, and an estimate of the eigenvector. There are, however, some caveats which will be discussed shortly.

The subroutine for determining eigenvalues by the power method may be demonstrated using program IV.33. Sample results are shown in listing IV.8. In the first example, $A(I,J)$ is the identity matrix, and thus there are multiple eigenvalues; all three equal unity. The reason is that in this case, the characteristic polynomial is $y(x) = (1 - x)(1 - x)(1 - x)$. The subroutine correctly determines the largest (and smallest) eigenvalue, but is confused with respect to the eigenvector; no individual term dominated.

In listing 8b, a diagonal matrix is considered which has five eigenvalues: 1, 2, 3, 4, and 5. The subroutine accurately determines the largest eigenvalue and comes close to finding the associated eigenvector (0,0,0,0,1). Another ten iterations would have reduced the fourth vector element from .0024 to .00026. Also, the returned eigenvalue would have been 4.9999999. In this case, the subroutine works quite well since there was a distinctly largest eigenvalue.

The third coefficient matrix considered (listing 8c) is arbitrary. The eigenvalue-seeking subroutine converges very quickly, indicating the existence of an eigenvalue considerably larger than any other.

The chief caution lies in the assumption that the vector returned by the subroutine is the eigenvector associated with the eigenvalue obtained. Also note that it is possible to have complex eigenvalues, even with a real number coefficient matrix. If the matrix is real and symmetric, all the roots of the characteristic polynomial must be real. However, if the matrix is simply real, complex conjugate pairs of roots may appear. The term *conjugate* means that if $a + ib$ is an eigenvalue, then so is $a - ib$. Obviously, the subroutine cannot return a complex eigenvalue.

The smallest eigenvalue of a matrix A can be obtained using the same subroutine by observing that the eigenvalues of the inverse of matrix A are the reciprocals of the eigenvalues of matrix A itself. Thus, the smallest eigenvalue can be found by first inverting A, and then applying the subroutine. The resulting eigenvalue is then the reciprocal of the value desired. If the matrix cannot be inverted, then the smallest eigenvalue is zero.

Eigenvalues and corresponding eigenvectors may be checked by recalling the defining eigenfunction equation and performing the matrix/vector multiplication.

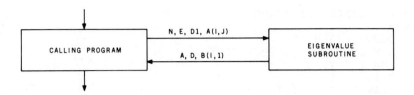

Figure IV.18. *Subroutine connection for determining the largest eigenvalue by the power method (EIGENPOW).*

Statements/Functions List

$+, -, *, /, <, >$, SQRT (or SQR)
ABS, IF/THEN, FOR/NEXT

Variables List

A, A(I,J), B, B(I,J), C(I,J), D, D1, E, I, M1
M2, N, N1, N2

Variables Passed to Subroutine

A(I,J), D1, E, N

Table IV.20 *Statements, functions, and variables used in the subroutine to determine the largest eigenvalue (EIGENPOW).*

Program IV.32:

```
42693 REM EIGENVALUE (POWER METHOD) SUBROUTINE (EIGENPOW)
42694 REM AX=LX
42695 REM A IS THE N X N MATRIX
42696 REM B IS AN ARBITRARY VECTOR
42697 REM E IS THE RELATIVE ERROR CHOSEN
42698 REM  D= COUNT OF THE NUMBER OF ITERATIONS
42699 REM SET PARAMETERS NEEDED FOR MULTIPLY SUBROUTINE
42700 M1=N
42701 N1=N
42702 M2=N
42703 N2=1
42704 REM GENERATE ARBITRARY NORMALIZED VECTOR B(I,1)
42705 FOR I=1 TO N
42706 B(I,1)=1/SQRT(N)
42707 NEXT I
42708 REM B = LAST EIGENVALUE ESTIMATE
42709 REM A = CURRENT EIGENVALUE ESTIMATE
42710 REM PICK AN INITIAL VALUE FOR THE EIGENVALUE GUESS
42711 B=1
42712 D=0
```

```
42713 REM START ITERATION
42714 A=0
42715 GOSUB 41900
42716 REM CONVERT C OUTPUT TO B
42717 FOR I=1 TO N
42718 B(I,1)=C(I,1)
42719 A=A+B(I,1)*B(I,1)
42720 NEXT I
42721 D=D+1
42722 A=SQRT(A)
42723 REM NORMALIZE VECTOR
42724 FOR I=1 TO N
42725 B(I,1)=B(I,1)/A
42726 NEXT I
42727 IF ABS((A-B)/A)<E THEN RETURN
42728 B=A
42729 IF D>D1 THEN RETURN
42730 GOTO 42714
```

Program IV.32: *Subroutine for determining the largest eigenvalue using the power method (EIGENPOW).*

```
REM PROGRAM TO DEMONSTRATE OBTAINING THE LARGEST EIGENVALUE OF A MATRIX
PRINT"WHAT IS THE SIZE OF THE MATRIX: ",
INPUT N
PRINT
DIM A(N+1,N+1),B(N+1,N+1),C(N+1,N+1)
FOR I=1 TO N
PRINT"INPUT ROW ",I
FOR J=1 TO N
INPUT A(I,J)
NEXT J
PRINT
NEXT I
PRINT
PRINT"INPUT LEVEL OF ACCURACY: ",
INPUT E
PRINT"INPUT NUMBER OF ITERATIONS AT WHICH TO STOP: ",
INPUT D1
REM FIND EIGENVALUE
GOSUB 42700
PRINT
PRINT"EIGENVALUE= ",A
PRINT
PRINT"EIGENVECTOR="
FOR I=1 TO N
PRINT B(I,1)
NEXT I
PRINT
PRINT"NUMBER OF ITERATIONS= ",D
END
```

Program IV.33: *Program to demonstrate obtaining the largest eigenvalue using program IV.32.*

```
WHAT IS THE SIZE OF THE MATRIX: ?3

INPUT ROW   1
?1
?0
?0

INPUT ROW   2
?0
?1
?0

INPUT ROW   3
?0
?0
?1


INPUT LEVEL OF ACCURACY: ?.0001
INPUT NUMBER OF ITERATIONS AT WHICH TO STOP: ?10

EIGENVALUE=  1

EIGENVECTOR=
 .57735027
 .57735027
 .57735027

NUMBER OF ITERATIONS=  1
```

Listing IV.8a: *Example of finding the largest eigenvalue for the identity matrix using program IV.33.*

```
WHAT IS THE SIZE OF THE MATRIX: ?5

INPUT ROW   1
?1
?0
?0
?0
?0

INPUT ROW   2
?0
?2
?0
?0
?0

INPUT ROW   3
?0
?0
?3
?0
?0

INPUT ROW   4
?0
?0
?0
?4
?0
```

Listing IV.8b:

```
INPUT ROW   5
?0
?0
?0
?0
?5


INPUT LEVEL OF ACCURACY: ?.000001
INPUT NUMBER OF ITERATIONS AT WHICH TO STOP: ?30

EIGENVALUE=  4.9999918

EIGENVECTOR=
 1.3421731E-19
 1.8014343E-11
 1.0234871E-06
 2.4178442E-03
 .99999708

NUMBER OF ITERATIONS=   27
```

Listing IV.8b: *Example of finding the largest eigenvalue of a diagonal matrix using program IV.33.*

Listing IV.8c:

```
WHAT IS THE SIZE OF THE MATRIX: ?6

INPUT ROW   1
?1
?4
?3
?5
?6
?7

INPUT ROW   2
?3
?1
?0
?7
?8
?5

INPUT ROW   3
?4
?6
?3
?2
?7
?6

INPUT ROW   4
?3
?4
?5
?2
?9
?2

INPUT ROW   5
?5
?2
?9
?9
?7
?8
```

```
INPUT ROW    6
?3
?4
?3
?2
?6
?4


INPUT LEVEL OF ACCURACY: ?.000000001
INPUT NUMBER OF ITERATIONS AT WHICH TO STOP: ?1000

EIGENVALUE=   28.63433

EIGENVECTOR=
.36981012
.35987121
.40173515
.38633046
.56545531
.32148277

NUMBER OF ITERATIONS=   9
```

Listing IV.8c: *Example of finding the largest eigenvalue of an arbitrary fifth order matrix using program IV.33.*

IV.10 Matrix Exponentiation and Differential Equations

In section IV.7 an approach to solving sets of simultaneous linear equations based on a matrix inversion subroutine was considered. In this section, simultaneous linear differential equations will be treated using the operation of matrix exponentiation.

Consider the situation in which the rates of change of the three variables X, Y, and Z are linearly related to the magnitudes of those variables. That is,

$$dX/dt = A(1,1) \times X + A(1,2) \times Y + A(1,3) \times Z$$
$$dY/dt = A(2,1) \times X + A(2,2) \times Y + A(2,3) \times Z \qquad \text{(IV.44)}$$
$$dZ/dt = A(3,1) \times X + A(3,2) \times Y + A(3,3) \times Z$$

For example, the variables might represent the number of customers in three restaurants. If one restaurant gets too crowded, customers may leave and go to the other two restaurants. The diagonal coefficients of matrix A would then be negative, and the off-diagonal elements would be positive.

The above set of equations may be written in vector form (reference 31):

$$dV/dt = A \times V \qquad \text{(IV.45)}$$

V is defined as:

$$\mathbf{V}(t) = \begin{bmatrix} X(t) \\ Y(t) \\ Z(t) \end{bmatrix} \qquad (IV.46)$$

The solution to this differential matrix equation is:

$$\mathbf{V}(t) = \exp(At) \times \mathbf{V}(0) \qquad (IV.47)$$

In this representation, **V**(0) is the initial state, and the exponential is defined through the MacLaurin series expansion:

$$\exp(At) = I + At + (At) \times (At)/2! + (At) \times (At) \times (At)/3! + \cdots \qquad (IV.48)$$

It is apparent that the key to the solution of the above set of linear differential equations is the matrix exponential, exp(At), and the key to that is matrix multiplication, an operation dealt with earlier. Shown in program IV.34 is a matrix exponentiation subroutine which performs the required function. As can be seen from figure IV.19, the exponentiation subroutine is largely built from other matrix subroutines presented in previous sections. This routine is certainly not fast, requiring more than 2 seconds per series term evaluated for a 3 by 3 matrix. Operation of this subroutine is demonstrated by program IV.35, with examples shown in listing IV.9.

The first example involves a zero dimensional matrix, or scalar; exp(1) = 2.718281828459045. The second example involves the identity matrix. In general, exp(XI) = I exp(X). In the third example, a matrix having all ones for elements is exponentiated with an interesting result: a matrix having all negative elements. This result is not correct and is the consequence of not including enough terms in the matrix series summation performed in the subroutine. The correct answer is shown in listing IV.9d.

Determining the number of terms required to obtain a given level of accuracy is complicated by the matrix multiplications. As evidenced by the third and fourth examples, the value of the determinant is of no use in estimating the convergence of the series, as DET(A) = 0 in those examples. Thus, we must either resort to the comparison of results for different series lengths or simply over-kill (use one hundred terms).

The subroutine given has a specific feature which might cause some difficulty for small values of X. In order to avoid overflow problems and to leave the matrix A unchanged upon return from the subroutine, zero values of X are avoided by replacing X = 0 with a small number (see line 42806). This overflow protection number can be made smaller if expansions involving very small values of X are desired.

The usual cautions regarding passing the appropriate variables to the subroutine also apply in this case. In addition, since the exponentiation subroutine calls several other subroutines to aid in performing the exponentiation operation, there is an extensive reserved variables list which must be kept in mind when using this program. See table IV.21, in addition to the other referenced tables.

This concludes the chapter on matrix operations. In Volume II the matrix operations developed in the previous sections will again be called upon for regression analysis.

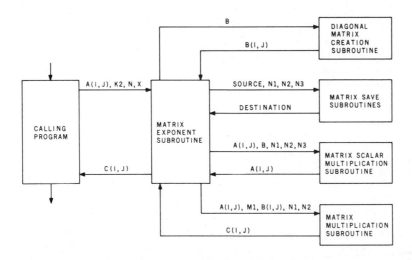

Figure IV.19. *Subroutine connection for the matrix exponentiation subroutine (MATEXP).*

Statements/Functions List

+, /, >
FOR/NEXT, GOSUB, IF/THEN, GOTO

Variables List

B, C(I,J), D(I,J), I, J, K1, K2, M1, M2, N, N1, N2, N3, X

Variables Passed to Main Subroutine

A(I,J), K2, N, X

Table IV.21 *Statements, functions, and variables used in the matrix exponentiation subroutine (MATEXP). This list is in addition to the items shown in tables IV.9, IV.11, IV.12, and IV.13.*

Program IV.34:

```
42796 REM MATRIX EXPONENT SUBROUTINE (MATEXP)
42797 REM INPUTS TO THE SUBROUTINE ARE THE MATRIX A, MATRIX
42798 REM SIZE N, NUMBER OF TERMS K2, AND VARIABLE X
42799 REM SET UP INDICES TO BE USED LATER
42800 N1=N
42801 N2=N
42802 N3=0
42803 M1=N
42804 M2=N
42805 REM GUARD AGAINST DIVIDE BY ZERO
42806 IF X=0 THEN X=.0000000000001
42807 REM INITIALIZE STORAGE MATRIX D(I,J)
42808 FOR I=1 TO N
42809 FOR J=1 TO N
42810 D(I,J)=0
42811 NEXT J
42812 NEXT I
42813 REM K2 IS THE NUMBER OF TERMS TO BE CALCULATED
42814 K1=0
42815 REM AUTO IDENTITY MATRIX IN B
42816 B=1
42817 GOSUB 42000
42818 REM MOVE B TO C
42819 GOSUB 42125
42820 REM ADD TO D
42821 GOSUB 42847
42822 K1=K1+1
42823 IF K1>=K2 THEN GOTO 42838
42824 REM SCALE MATRIX A BY X/K1
42825 B=X/K1
42826 GOSUB 42200
42827 REM MULTIPLY A TIMES B
42828 GOSUB 41900
42829 REM ADD RESULT TO MATRIX D
42830 GOSUB 42847
42831 REM MOVE C TO B
42832 GOSUB 42100
42833 REM RETURN MATRIX A TO ORIGINAL CONDITION
42834 B=K1/X
42835 GOSUB 42200
42836 REM CONTINUE SUMMATION
42837 GOTO 42822
42838 REM MOVE RESULT IN D TO C
42839 FOR I=1 TO N
42840 FOR J=1 TO N
42841 C(I,J)=D(I,J)
42842 NEXT J
42843 NEXT I
42844 REM RETURN TO CALLING PROGRAM
42845 RETURN
42846 REM D(I,J) IS USED FOR TEMPORARY STORAGE
42847 FOR I=1 TO N
42848 FOR J=1 TO N
42849 D(I,J)=D(I,J)+C(I,J)
```

```
42850 NEXT J
42851 NEXT I
42852 RETURN
```

Program IV.34: *Matrix exponentiation subroutine (MATEXP).*

```
REM PROGRAM TO DEMONSTRATE THE MATRIX EXPONENT SUBROUTINE
PRINT"INPUT THE SIZE OF THE MATRIX TO BE EXPONENTIATED",
INPUT N
DIM A(N+1,N+1), B(N+1,N+1), C(N+1,N+1), D(N+1,N+1)
FOR I=1 TO N
PRINT"INPUT ROW ",I
FOR J=1 TO N
INPUT A(I,J)
NEXT J
PRINT
NEXT I
PRINT
REM PRINT MATRIX A
N1=N
N2=N
N3=0
M=N
PRINT"MATRIX A="
PRINT
REM MOVE A TO C
GOSUB 42150
REM PRINT
GOSUB 1020
PRINT"INPUT THE VARIABLE X",
INPUT X
PRINT
PRINT"INPUT THE NUMBER OF TERMS TO BE CALCULATED",
INPUT K2
REM OBTAIN EXPONENT
GOSUB 42800
PRINT
PRINT"EXP(AX)= "
PRINT
GOSUB 1020
PRINT
END
```

Program IV.35: *Program to demonstrate the matrix exponentiation subroutine shown in program IV.34.*

```
INPUT THE SIZE OF THE MATRIX TO BE EXPONENTIATED?1
INPUT ROW   1
?1

MATRIX A=

        1

INPUT THE VARIABLE X?1

INPUT THE NUMBER OF TERMS TO BE CALCULATED?10

EXP(AX)=

        2.72
```

Listing IV.9a: *Results of using the matrix exponentiation subroutine (MATEXP) to determine exp(1).*

```
INPUT THE SIZE OF THE MATRIX TO BE EXPONENTIATED?3
INPUT ROW   1
?1
?0
?0

INPUT ROW   2
?0
?1
?0

INPUT ROW   3
?0
?0
?1

MATRIX A=

        1    0    0
        0    1    0
        0    0    1

INPUT THE VARIABLE X?1

INPUT THE NUMBER OF TERMS TO BE CALCULATED?10

EXP(AX)=

        2.72  0     0
        0     2.72  0
        0     0     2.72
```

Listing IV.9b: *Results obtained from exponentiating the identity matrix; exp(I).*

```
INPUT THE SIZE OF THE MATRIX TO BE EXPONENTIATED?3
INPUT ROW   1
?1
?1
?1

INPUT ROW   2
?1
?1
?1

INPUT ROW   3
?1
?1
?1

MATRIX A=

         1     1     1
         1     1     1
         1     1     1

INPUT THE VARIABLE X?-2

INPUT THE NUMBER OF TERMS TO BE CALCULATED?10

EXP(AX)=

       -2.89 -3.89 -3.89
       -3.89 -2.89 -3.89
       -3.89 -3.89 -2.89
```

Listing IV.9c: *Results obtained from exponentiating twice the matrix A; exp(−2A). Only ten terms were summed, which is not sufficient as may be seen from listing IV.9d.*

Listing IV.9d:
```
INPUT THE SIZE OF THE MATRIX TO BE EXPONENTIATED?3
INPUT ROW   1
?1
?1
?1

INPUT ROW   2
?1
?1
?1

INPUT ROW   3
?1
?1
?1

MATRIX A=

         1     1     1
         1     1     1
         1     1     1

INPUT THE VARIABLE X?-2
```

```
INPUT THE NUMBER OF TERMS TO BE CALCULATED?100
EXP(AX)=

        .67   -.33  -.33
       -.33    .67  -.33
       -.33   -.33   .67
```

Listing IV.9d: *Rerun of the case exp(−2A), but summing 100 terms.*

CHAPTER V

RANDOM NUMBER GENERATORS

One of the most widely used mathematical concepts in computing is the random number generator. It is employed both to simulate the chance associated with physical processes and to provide a mathematical tool for analyzing otherwise intractable problems. In the chance category, random number generators are used to simulate traffic, war, business, and other systems which have an unpredictable nature, and are heavily relied upon in Monte Carlo analysis. Random number generators can also be used as mathematical devices to perform integrations of complex multidimensional problems.

Monte Carlo analysis represents one of the most important uses for random number generators because the methodology has relevance to both science and industry. In industry, for example, the output of a manufacturing process is the result of many input parameters. Characteristics of the output, such as yield and quality, fluctuate because of variations in the input parameters. Even if the entire process were modeled exactly and the distributions of the input variables were known, it might be impossible to derive an analytical solution which would relate the output variations to the parameters, describing the distributions of the individual inputs. A brute force computer evaluation might be attempted, but the size of the matrix of calculations could be prohibitive. For example, ten values from each of fifty input distributions might be selected, giving 10^{50} possible I/O combinations. Monte Carlo analysis offers a different, more time-efficient approach in which a random number generator, following the characteristics of a particular distribution, is called to supply an input value. These values are then used to calculate a single output for the process. This procedure may be repeated several hundred times to develop a rough statistical picture of the overall behavior of the process using less computer time than the matrix approach. The analyst can then tighten the control (eg: the standard deviation) on a particular input variable and determine if the output distribution is significantly affected. A set of process specifications can then be developed by factoring in the cost of the controls and the improved return on the process.

It is apparent from the preceding paragraph that many types of random number generators are required since the input distributions may be quite varied. In this chapter, several random number generators are provided, all based on the uniform distribution generator. This type of generator is important for several reasons. First, it produces a continuum of numbers. Second, it is relatively easily approximated. Third, it can be readily used to create other generators.

The random number generators given in the next several sections are derived in three general ways. One technique employs the Central Limit Theorem in mathematics to directly approximate a Normal (or shifted Gaussian) distribution. Another technique, exemplified by the derivation of the Binomial random number generator, uses the uniform distribution to simulate the random selection of objects, such as used in determining simple probabilities. The third method, repeatedly employed in this book, is based on analytically integrating the *probability density function* for two generators, one being the BASIC RND function and the other the desired function. The result is two *cumulative distribution functions*. These two functions are then set equal. Some algebraic manipulation is then performed, and the desired random number simply becomes a transformation of the value obtained from the RND function. This will all be more fully discussed in later sections.

The above techniques are used to provide the random number generator subroutines listed in table V.1. They all execute slower than the RND functions available in most BASIC interpreters because they are not written in machine language and because all but the linear generator call the RND function at least once in the process of generation. For those computer systems which do not have a uniform distribution random number generator, the subroutine called LINEAR is supplied as a replacement.

Following the format of previous sections, tables are provided which describe the statements, functions, and variables used. Test demonstrations are also given.

Generator	Bytes	Parameters	Execution Time (sec)
Linear	374	U=0.5, V=1	0.19
Normal	179	U=1, V=1	1.41
Poisson	211	U=10	0.61
Binomial	169	B=0.1, N=100	2.9
Exponential	90	U=1	0.16
Fermi	230	U=1, V=0.4	7.2 (V=0.1 is 10×faster)
Cauchy	117	U=1	0.16
Gamma	176	B=1	1.35
Beta	352	A=1, B=1	0.09
		A=2, B=1	0.56
Weibull	120	U=0.5, V=2	0.38

Table V.1 *Random number generator subroutine memory requirements and execution times. For comparison, the RND function in North Star BASIC consumes 0.033 seconds per number generated. The parameters passed to the subroutine are shown in the third column.*

V.1 The Uniform Distribution, RND

In many high-level languages a uniform distribution random number generator is provided in the same manner as are the trigonometric functions (see references 11, 15, and 20). In BASIC, this function generally has the form RND(x), where the argument, x, provides some control over what the function does. Unfortunately, the control function varies among different interpreters and compilers. For example, in North Star BASIC, an argument between 0 and 1 restarts the generator using the value provided as a *seed*, or starting point. RND(0) then calls the next pseudorandom number in the sequence. In Microsoft BASIC (MITS 8 K), a negative argument denotes a new seed, RND(0) returns the last number supplied, and x between 0 and 1 returns the next number in the sequence. Thus, there are obvious incompatibilities between BASIC interpreters with respect to this function.

The RND function returns a random number in the range 0 to 1. The distribution is called uniform since each number in that range is equally likely to occur. In the following explorations of the RND functions, the North Star version is assumed.

It is useful at this point to introduce the concept of the *probability density function*, $p(x)$, which for the uniform distribution is:

$$p(x) = 1 \qquad \text{for } 0 < x < 1 \qquad \text{(V.1)}$$
$$p(x) = 0 \qquad \text{otherwise}$$

The meaning of $p(x)$ is simple. The probability that the random number which is generated will fall between x and $x + dx$ is $p(x)dx$, where dx is the incremental range. For those acquainted with calculus, the differential form just used can be converted into an integral equation:

$$P(a < x < b) = \int_a^b p(x)\, dx \qquad \text{(V.2)}$$

P is the probability that the random number which is returned will fall between a and b. (What is the probability that the number returned will be between 0 and 0.2? Answer: 0.2.)

The uniform random number generator can be easily used for some classical problems. For example, to simulate the roll of an unweighted die, use the following simple routine:

```
X = RND(0)
N = INT(6*X) + 1
```

The first statement creates a random number, X. The second statement finds into which one of the six equal segments in the range 0 to 1 that X has fallen, each segment corresponding to a numbered face on the die. For example, if $X = 0.3, N = 2$ would be returned. N (which is an integer) can range between 1 and 6 inclusively, with each value equally likely. (This occurs because the INT function, when operating on positive numbers, rounds off the argument to the next lowest integer.) To add some weighting (loading) to the die, a more complicated program which uses unequally sized intervals can be written.

The idea of using the continuous uniform distribution to create discrete random numbers can be applied to test the generator. Program V.1 shows a routine (not a library subroutine) which calls upon the RND function 1000 times and divides the results into ten groups (dectiles) corresponding to the range 0 to 0.1, 0.1 to 0.2, and so on. The program then uses the plotting subroutine discussed earlier to display the distribution of results (see figure V.1). The X axis ranges from 0 to 2, and therefore contains twenty dectiles. Only the dectiles in the range of 0 to 1 are expected to have values other than 0 since the RND function returns numbers only between those limits. The largest number observed in any dectile was 115, and that occured in the fifth. Observe that a roughly equal number fell into each dectile (there is some scatter due to the finite size of the sample; a ten-unit deviation is expected), with no values observed beyond $X = 1$. The probability density function may be graphically displayed as shown in figure V.2. This may be compared with the 1000-number sample results in the following manner.

We may estimate the probability distribution function that generated the samples by using the samples themselves. The number observed to fall in dectile i is:

$$N(i) \approx N \, p(i) \, dx(i) \qquad (V.3)$$

N is the total number of trials, 1000; $p(i)$ is the probability density function in that dectile (ideally unity); and $dx(i)$ is the length of the interval (.1). Thus, we have the estimate:

$$p(i) \approx N(i)/[N \, p(i) \, dx(i)] = N(i)/100 \qquad (V.4)$$

For example, if $N(i) = 115$, the maximum observed, then the estimate of $p(i)$ in that interval is 1.15. Transforming the results shown in figure V.1 in this manner, we have the points shown in figure V.2. The correspondence is reasonably accurate. If 10,000 numbers had been generated instead, the fit would have been even better. The distribution estimated from the numbers generated is called the *sample distribution*. The solid curve shown in figure V.2 is called the *population distribution*. The sample distribution more closely approximates the population distribution as the number of values generated increases.

At this point, a word of caution is necessary. No random number generator is perfect. Thus, the RND function when applied endlessly may not lead to a perfectly uniform and uncorrelated sequence. However, as shown in the next section, the North Star generator appears to be adequate with respect to its *moments*.

The range of numbers generated can be easily altered with one statement. If U is the desired center for the distribution and V is the spread, then the following statement transforms RND to the new distribution:

$$Y = U + V*(RND(0) - 0.5)$$

If a square (ie: two-dimensional) uniform distribution in X and Y is desired, then the statements to be used are:

$$X = RND(0)$$
$$Y = RND(0)$$

The discussion in this section and the introduction to the chapter have been somewhat lengthy in order to introduce the concepts employed in the following sections. These sections are relatively brief and are only meant to give an outline of the techniques that can be used to derive the distributions and create the resulting subroutines. The random number generator subroutines can be called without knowledge of how they were derived, and the reader can simply rely on the tables and subroutine-connection guides for implementation.

```
REM PROGRAM TO DEMONSTRATE THE USE OF THE UNIFORM
REM RANDOM NUMBER GENERATOR, RND(0)
L=80
REM SET PLOT WIDTH
N=21
DIM C(22),D(22),E(22)
FOR I=1 TO 21
D(I)=0
C(I)=(I-1)/10
NEXT I
REM SET SEED
Z2=RND(.5)
FOR I2=1 TO 1000
Z2=10*RND(0)+1
Z2=INT(Z2)
D(Z2)=D(Z2)+1
NEXT I2
PRINT"DISTRIBUTION OF THE 1000 RANDOM NUMBERS"
PRINT"IN TERMS OF THE NUMBER PER INTERVAL OF LENGTH 1/10"
PRINT
PRINT
GOSUB 40100
PRINT
PRINT
END
```

Program V.1: *Program to demonstrate the use of the RND function.*

Figure V.1:

```
RUN

DISTRIBUTION OF THE 1000 RANDOM NUMBERS
IN TERMS OF THE NUMBER PER INTERVAL OF LENGTH 1/10

***** DATA PLOT (SCALED) *****

MIN ORDINATE=   0    MAX ORDINATE=  115
INITIAL ABSCISSA VALUE=  0
```

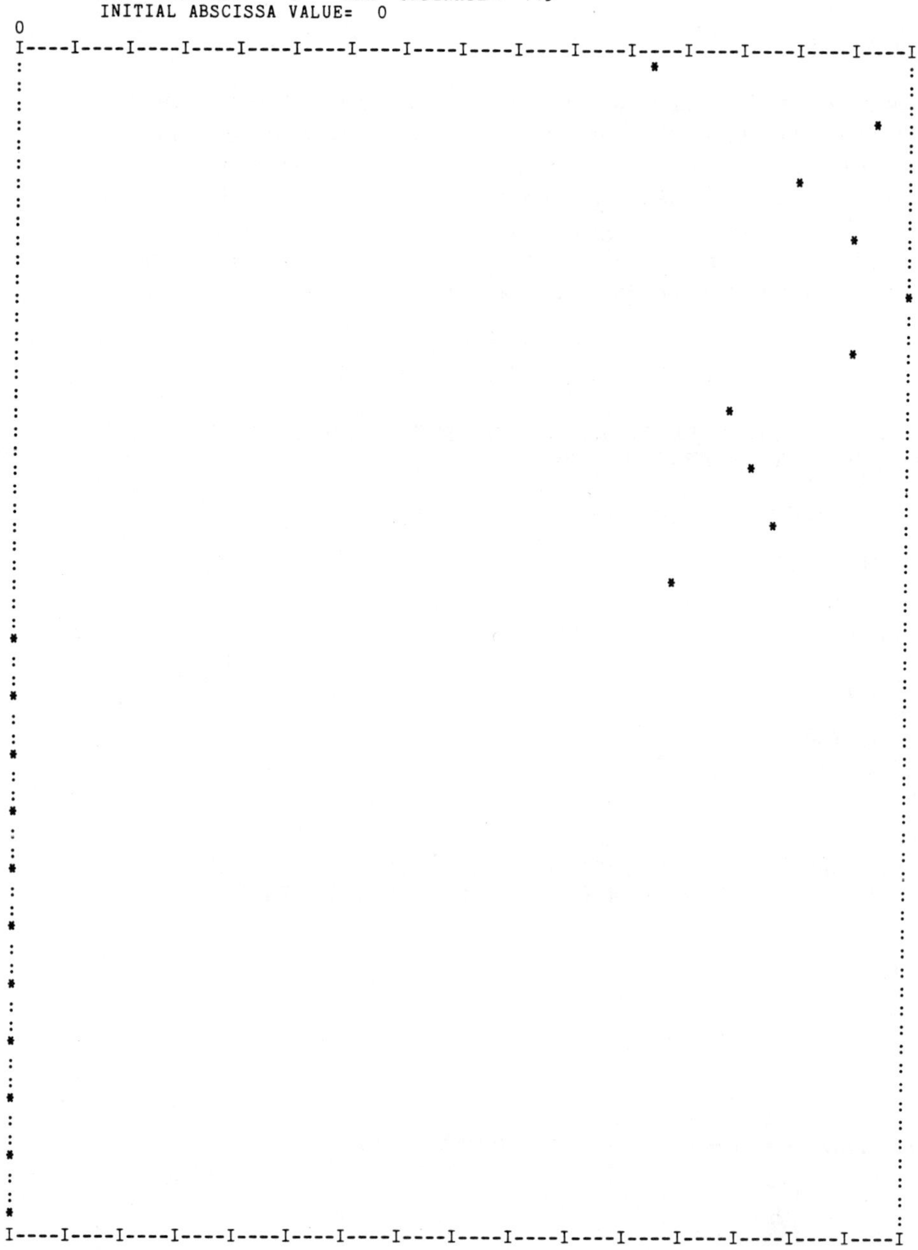

END ABSCISSA VALUE= 2

READY

Figure V.1. *Listing resulting from running program V.1. Dectile distribution of a 1000-number long sequence generated by RND(0).*

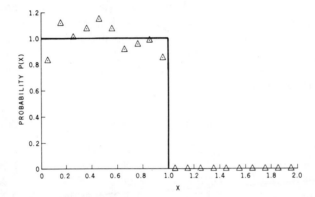

Figure V.2. *RND(0) distribution (solid lines) and sample results. Sample results taken from figure V.1.*

V.2 The Linear Distribution

The linear distribution referred to in this section is conceptually the same as the uniform distribution. A different name is used to distinguish the BASIC subroutine (see program V.2) from the RND function.

The subroutine shown in program V.2 is empirical, but follows a commonly used mathematical strategy in which a multi-digit number is generated in some pseudorandom fashion, and then only the trailing digits are used. If the value to the right of the decimal point is chosen as the result, the range of the numbers generated will be 0 to 1. The technique used in this case is to take a seed, D, transform it into a number greater than unity, divide this value by the irrational number e, add 1 to this result, and multiply the sum by the irrational number, which is $\sqrt{2}$. This value is subsequently added to the previous random number in the sequence and multiplied by an increasing integer. Multiples of π and unity are then subtracted to bring the near-final value into the range 0 to 1. Even after all these magical steps are performed, the resulting distribution of numbers is not entirely uniform, and an empirical correction is performed. Finally, as discussed in the previous section, the pseudorandom numbers generated are scaled and shifted so that they are centered about U with a spread of V.

The subroutine connection diagram is shown in figure V.3. D is the seed, U is the desired mean, and V is the desired range. As in the previous section, a 1000-number long sequence can be generated, divided into segments, and compared against the expected distribution. The results appear in figure V.4; they are similar to those appearing in figure V.2. Thus, at first glance, the RND and the subroutine appear equivalent. They may be compared as follows.

One of the key theorems in statistics states that a random number distribution can be completely described by its *moments*. The North Star RND and the empirical generators were tested by creating ten sets of 1000-number long sequences. The moments calculated are:

Moment	Ideal Pop.	North Star RND	Empirical
First (mean)	0.500	$0.497 \pm .004$	$0.494 \pm .011$
Second (about mean)	0.0833	$0.083 \pm .002$	$0.081 \pm .002$
Third (about mean)	0.0000	$-.0002 \pm .0003$	$-.001 \pm .001$
Fourth (about mean)	0.0125	$0.0123 \pm .0004$	$0.0118 \pm .0004$

The above results indicate that both the RND and empirical generators are reasonable by these tests. Chi-square tests also support this finding.

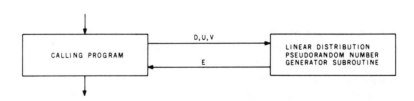

Figure V.3. *Subroutine connection for the linear distribution pseudorandom number generator (LINEAR).*

Statements/Functions List

+, −, *, /, >
ABS, IF/THEN, INT

Variables List

A, B, C, D, E, I9, U, V

Variables Passed to Subroutine

D, U, V

Table V.2: *Statements, functions, and variables used in the linear pseudorandom number generator subroutine (LINEAR). Note that I9 is automatically incremented to create the next number in the sequence. Care should be exercised in using this variable elsewhere.*

```
42899 REM LINEAR RANDOM NUMBER GENERATOR (LINEAR)
42900 REM U=MEAN, V=SPREAD, D=SEED
42901 I9=I9+1
42902 A=3.14159265358979323846
42903 B=2.71828182845904523536
42904 C=1.41421356237309504880
42905 D=1+ABS(D)
42906 E=E+(1+D/B)*C
42907 E=E*I9
42908 E=E-A*INT(E/A)
42909 E=E-INT(E)+.018
42910 IF E>.1 THEN E=E+.009
42911 IF E>.2 THEN E=E-.002
42912 IF E>.3 THEN E=E-.005
42913 IF E>.4 THEN E=E-.005
42914 IF E>.5 THEN E=E-.015
42915 E=V*(E-.5)+U
42916 RETURN
```

Program V.2: *Linear distribution random number generator subroutine which has been derived semi-empirically. The shaping of the distribution is accomplished in statements 42909 thru 42914.*

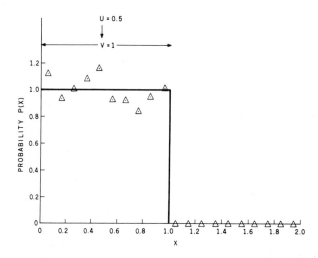

Figure V.4. *Expected distribution for the empirical uniform distribution random number generator with some sample results.*

V.3 The Normal Distribution

One of the most widely employed probability distribution functions is the *Normal distribution*, or *shifted Gaussian*. Its density function has the form:

$$p(x) = (V\sqrt{2\pi})^{-1} \exp[-(x-U)^2/(2V^2)] \quad \text{for } -\infty < x < +\infty \quad (V.5)$$

where U = the mean
       V = the standard deviation

This distribution has very fundamental properties that are detailed in any standard statistics text (see references 14, 32, or 34). It can be approximated as follows. According the the Central Limit Theorem (see reference 33), if a large group of independent random variables is summed, with each variable taken from the same distribution, then the sum itself is a random variable. This random variable has a distribution which approximates the Normal distribution, the approximation becoming better as the number of variables summed increases. This powerful result can be used to intuitively explain why the deviations observed in many engineering, physical, and psychological processes can often be fitted to a Normal probability density function. Even other distributions, such as the Binomial and Chi-Square, approach the Normal distribution in the limit of large numbers, giving the characteristic bell-shaped curve shown in figure V.6.

The Central Limit Theorem can be used to generate normally distributed random numbers by repeatedly summing random numbers derived from the uniform distribution. The mathematical form for this (see reference 29) is:

$$y = \sum_{i=1}^{k} \frac{x - 0.5}{\sqrt{k/12}} \qquad (V.6)$$

In the above summation, x is the random variable derived from the RND function, and y is the resultant Gaussian distributed random number. The reference suggests that $k = 12$ be used. Although a good choice for reasonably accurate calculations, this falls short when there is interest in the low probability tails of the distribution, as in the case of data transmission error analysis. For $k = 12$, the maximum value of y is 6. Beyond this point the approximation, $p'(y)$, is zero while the correct value, $p(y)$, is actually non-zero. For example, $p(6) \cong 2 \times 10^{-9}$. For low probability analysis, a higher value of k is required. The subroutine in program V.3 uses $k = 48$, which is more than sufficient for any conceivable situation; $p(24) \cong 10^{-126}$! The approximation is thus very good, providing that the uniform-distribution random number generator used is truly random. Note that it need not be really uniform, as that restriction is not required by the Central Limit Theorem.

The summation shown above leads to a normalized Gaussian distribution where $U = 0$ and $V = 1$. It can be easily changed to a general Normal distribution of mean U and standard deviation V by use of the transformation:

$$y' = U + Vy$$

This is included in the subroutine shown in program V.3. The subroutine connection is described in figure V.5. As was done in previous sections, the subroutine was partially tested by using it to generate a 1000-number long sequence and plotting the segmented results (see figure V.6).

The normally distributed random number generator can be easily extended to another distribution called the Log-Normal:

$$p(z) = \frac{\exp[-(\ln(z) - U)^2/(2V^2)]}{zV\sqrt{2\pi}} \qquad (V.7)$$

The Log-Normal distribution function describes a random variable, z, whose logarithm is normally distributed with mean U and standard deviation V (see reference 13). This type of distribution is often observed in powder technology, in which the measured particle sizes appear to be log-normally distributed, giving a relatively high probability of finding both "fines" and "boulders." The parameters which go with this distribution are:

- mean of $z = \exp(U + V^2/2)$
- standard deviation of $z = [\exp(V^2) - 1]\exp(V^2 + 2U)$
- median of $z = \exp(U)$ (V.8)
- mode of $z = \exp(U - V^2)$

A log-normally distributed sequence of random numbers can be generated using the random numbers generated for the normal distribution as follows. A random number, x, is generated, belonging to the Normal distribution having a mean U and standard deviation V. Since this represents $\ln(z)$, we have $x = \ln(z)$, or:

$$z = \exp(x) \qquad (V.9)$$

Observe that although x may be positive or negative, z is always positive, as might be expected for physical variables such as particle sizes.

The extension of the normal generator to the log-normal generator is simple. An overflow can occur if the argument in the exponent is too large, but this is highly unlikely if the random numbers generated are to represent physical processes. That is, values of U and V corresponding to real situations should not lead to an overflow.

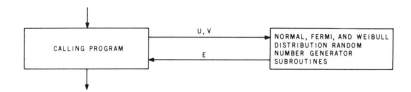

Figure V.5. *Subroutine connection for the Normal, Fermi and Weibull distribution random number generators.*

Statements/Functions List

+, −, *, /
FOR/NEXT, RND

Variables List

E, 19, U, V

Variables Passed to Subroutine

U, V

Table V.3 *Statements, functions, and variables used by the Normal-distribution random number generator subroutine (NORMAL).*

```
42923 REM NORMAL DISTRIBUTION BY CENTRAL LIMIT THEOREM (NORMAL)
42924 REM U=MEAN, V=STANDARD DEVIATION, E=RANDOM NO. GENERATED
42925 E=0
42926 FOR I9=1 TO 48
42927 E=E+RND(0)-.5
42928 NEXT I9
42929 E=V*E/2+U
42930 RETURN
```

Program V.3: *Normally distributed random number generator subroutine (NORMAL) based on the use of RND.*

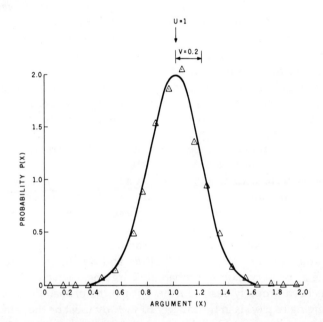

Figure V.6. *Statistical distribution for the Normal random number generator (NORMAL), with sample results for $U = 1$; $V = 0.2$.*

V.4 The Poisson Distribution

Another important distribution which is often observed in physical and statistical processes is the Poisson:

$$p(x) = U^x \exp(-U)/x! \tag{V.10}$$

where

$$U > 0$$
$$x = 0, 1, 2, 3, \ldots$$

The characteristic parameter in this distribution is U, the mean. The standard deviation is represented by $\sqrt{U}$. One of the special characteristics of the Poisson distribution is that it is not continuous and is defined only at non-negative integral values of x. The Binomial distribution (to be discussed later) shares this property, and, in the limit of large numbers and low probability, the Binomial distribution approaches the Poisson.

The parameter U can be visualized as the average number of events occurring per unit time. The probability of observing a given number of events, x, is p(x). It is possible, though unlikely, to have no event, highly likely to have many events near the mean, and somewhat unlikely to have an observed number of events much greater than the mean. This distribution is very useful in describing situations such as the incoming call rate at a telephone exchange, the frequency of radioactive counts in a Geiger tube, and similar, independent, random occurrences.

Figure V.8 shows the shape of the Poisson distribution for the case U = 6. A smooth curve was drawn through the distribution values (circles) for the sake of visualization. The random number generator to be presented is capable of generating a continuum of values or a discrete set.

The basic concept used to derive the Poisson random number generator involves the cumulative probability distribution function discussed in section V.1. Since this approach is employed in several following sections, an extended discussion is appropriate.

To derive the generating function we proceed as follows. If p(x) is the desired distribution, and p'(y) the uniform distribution, and we wish to relate p(x) to p'(y), we can invoke a "conservation of probability" argument which takes the following form. If x has been properly derived from y, then the probability that the random variable belonging to p'(y) is in the range y to y + dy must be the same as the probability that the random variable belonging to p(x) is in the range x to x + dx. For example, if we had stretched a uniform distribution having U = 0.5 and V = 1, to one having U = 1 and V = 2, then x = 2y and dx = 2dy. In differential equation form this concept may be expressed as:

$$p(x)\, dx = p'(y)\, dy \qquad (V.11)$$

If both distributions start from zero, the above equation can then be integrated to give:

$$\int_0^X p(x)\, dx = \int_0^Y p'(y)\, dy \qquad (V.12)$$

The integrals are the cumulative distribution functions, P and P':

$$P(x < X) = P'(y < Y) \qquad (V.13)$$

Since we know the analytic forms of both p(x) and p'(y), the relationship between x and y is in principle established. Because p'(y) = 1 (the uniform distribution), the above equation simplifies to:

$$P(x < X) = Y \qquad (V.14)$$

The first objective of the random number generator subroutine is to use the RND function, or optionally, the empirical generator described in section V.2, to obtain a value for Y. The subroutine then must locate, by some means, a value for X which satisfies the cumulative distribution relationship. In the case of the Poisson distribution, which is discrete, the goal could be to find the value of X that most closely satisfies the following equation:

$$Y = \sum_{x=0}^{X} U^x \exp(-U)/x! \qquad (V.15)$$

The subroutine shown in program V.4 linearly interpolates between the two integral X values which bracket the equality, thus forming a piecewise approximation to the continuous curve in figure V.8. As the mean becomes larger, the fit becomes smoother.

To test the generator, as in the previous sections, a 1000-number long sequence was generated, segmented, and plotted as shown in figure V.8. The generated values are represented by the triangles.

The Poisson random number generator produces a continuum of values, while the true Poisson distribution is defined only at the integer values of X. If integers are desired, the following statement should be executed after the random number E has been returned by the subroutine:

$$E = INT(E + 0.5)$$

There are two basic precautions involved in using the Poisson random number generator subroutine. The first is that it does not fit the distribution well for low values of U. However, from figure V.8 it is apparent that U = 6 is certainly large enough. The second precaution involves overflow, which may occur if U is too large. In North Star BASIC (Release 3), for example, U can be as large as 145.

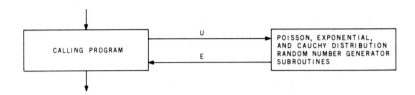

Figure V.7. *Subroutine connection for the Poisson, exponential, and Cauchy distribution random number generators.*

Statements/Functions List

$+, -, *, /, >$
EXP, GOTO, IF/THEN, RND

Variables List

E, X, X1, Y, Y1, U

Variables Passed to Subroutine

U

Table V.4 *Statements, functions, and variables used by the Poisson-distribution random number generator subroutine (POISSON).*

```
42948 REM POISSON RANDOM NUMBER GENERATOR (POISSON)
42949 REM INPUT PARAMETER U
42950 X=RND(0)*EXP(U)
42951 X1=1
42952 Y1=1
42953 Y=0
42954 IF X1>X THEN GOTO 42959
42955 Y=Y+1
42956 Y1=Y1*U/Y
42957 X1=X1+Y1
42958 GOTO 42954
42959 IF Y>0 THEN Y=Y-(X1-X)/Y1
42960 E=Y
42961 RETURN
```

Program V.4: *Poisson distributed random number generator subroutine (POISSON) using the RND function.*

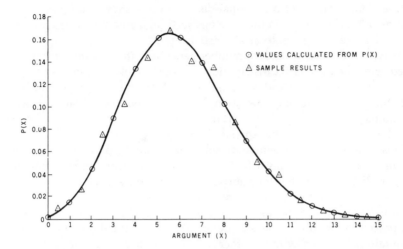

Figure V.8. *Statistical distribution for the Poisson random number generator (POISSON), with some sample results for U = 6.*

V.5 The Binomial Distribution

One of the first distributions considered in elementary courses on statistics is the binomial. It is often introduced in the example form of determining the probability of drawing black balls from an urn containing both black and white balls, when, after each selection is made, the chosen ball is returned to the urn. If K is the number of black balls picked after N selections, and if B is the fraction of black balls in the urn, then the probability density distribution function which describes the probable outcome is:

$$p(K) = \frac{N! \; B^K (1 - B)^{N-K}}{K! \, (N - K)!} \qquad \text{(V.16)}$$

for K = 0, 1, 2, 3, 4, . . . , N; 0 ≤ B ≤ 1

The parameters for this distribution are:

$$\text{mean} = NB \qquad \text{(V.17)}$$
$$\text{standard deviation} = \sqrt{NB(1 - B)}$$

As with the Poisson distribution, the binomial distribution is defined only at integral values of K. If the probability of a successful event per trial is B, then in N trials the probability of K events occurring is p(K). Events only come in integral quantities. An example of a binomial distribution is shown in figure V.10 for the case B = 0.9, N = 15. As in the previous section, a smooth curve has been drawn through the discrete values to aid in visualization.

As with the Poisson random number generator, the binomial generator can be derived using the cumulative distribution approach. However, in this case, there is a much simpler method. Instead, we simulate the N trials by calling on the RND function N times and tabulating how often the random number obtained is less than B (for example, picking only black balls from the urn containing both white and black balls). The subroutine for this appears in program V.5.

This random number generator subroutine can be partially tested, as in the previous sections, by creating a 1000-number long sequence and plotting the results (see figure V.10). Note that the subroutine returns only integers.

There are several precautions involved in using the binomial random number generator subroutine. First, as can be seen from figure V.9, the probability parameter passed to the subroutine is B, not P (the conventional descriptor). B was used to maintain consistency with the other random number generator subroutines given in this chapter. Second, the distribution which is generated is based on the uniform distribution random number generator and will be affected by any distortion in that distribution. Third, the binomial random number generator is slow, as evidenced by table V.1. The execution time is proportional to N. For N = 100, the execution time is approximately 3 seconds; for N = 1000, this would increase to about 30 seconds. To circumvent this problem, an asymptotic property of the binomial distribution can be called upon. As N gets large (N > 20), the binomial distribution approaches the shape of the Poisson distribution, provided that B is reasonably small (say B < 0.2). In that case, the Poisson distribution may be used with U = NB. The restriction on B is largely due to the fact that the binomial distribution is defined only on the closed interval $0 \le K < N$. The Poisson distribution applies to the open interval $0 \le K < \infty$.

If B is small enough, the Poisson random number generator using U = NB will have a low probability of creating a number larger than N, a possibility not allowed by the binomial distribution. To summarize, the main source of concern in using the Poisson distribution to approximate the binomial is the possible occurrence of numbers greater than N.

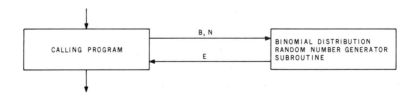

Figure V.9. *Subroutine connection for the Binomial random number generator (BINOMIAL).*

Statements/Functions List

+, <
FOR/NEXT, IF/THEN, RND

Variables List

B, E, K, N, Y1

Variables Passed to Subroutine

B, N

Table V.5 *Statements, functions, and variables used by the Binomial-distribution random number generator subroutine (BINOMIAL).*

```
42972 REM BINOMIAL RANDOM NUMBER GENERATOR (BINOMIAL)
42973 REM B=PROBABILITY PER TRIAL
42974 REM N=NUMBER OF TRIALS
42975 E=0
42976 FOR K=1 TO N
42977 Y1=RND(0)
42978 IF Y1<B THEN E=E+1
42979 NEXT K
42980 RETURN
```

Program V.5: *Binomially distributed random number generator subroutine (BINOMIAL) using the RND function.*

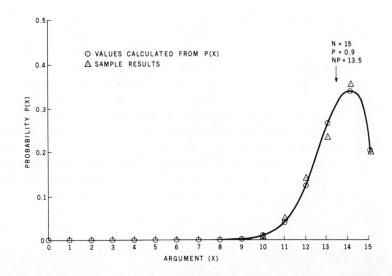

Figure VI.10. *Statistical distribution for the Binomial random number generator (BINOMIAL), with same results for $N = 15$; $B = 0.9$.*

V.6 The Exponential Distribution

In this section we consider the exponential distribution, which has the form:

$$p(x) = (1/B) \exp(-x/B) \qquad \text{for } 0 \leq x \qquad (V.18)$$

The parameters for this distribution are:

$$\text{mean} = B \qquad (V.19)$$

$$\text{standard deviation} = B$$

An example of a process in which such a distribution can be used is telephone switchboard simulation. The probability that no telephone calls occur within a given time, x, when the average call rate is B, can be approximated by the exponential distribution (see references 26 and 34). For large time intervals, there is a very low probability that no calls will occur; for short time intervals there is a high probability. The shape of the distribution, which is continuous, is shown in figure V.11.

To devise an appropriate subroutine we call upon the cumulative distribution approach discussed in section V.4. If Y is the random number obtained from the uniform distribution random number generator, and X is the desired exponentially distributed variable, we have:

$$Y = 1 - \exp(-X/B) \qquad (V.20)$$

This can be converted to:

$$X = -B \ln(1 - Y) \qquad (V.21)$$

Observe that when $Y = 0$, $X = 0$, and when $Y = 1$, $X = $ infinity.

The above transformation is used in the exponential random number generator subroutine in program V.6. The subroutine was tested by creating a 1000-number long sequence, segmenting it, and plotting (see figure V.11).

A potential problem in using this subroutine involves the logarithm function. If the number generated by the RND function is very close to unity, an overflow can occur. Highly unlikely, this occurrence is not guarded against in the subroutine. A problem might occur in a situation in which the RND function used is not uniform, thus possibly supplying too many numbers near unity.

Statements/Functions List

—, *

LOG, RND

Variables List

E, U, X

Variables Passed to Subroutine

U

Table V.6 *Statements, functions, and variables used by the exponential-distribution random number generator subroutine (EXPONENT).*

```
42998 REM EXPONENTIAL RANDOM NUMBER GENERATOR (EXPONENT)
42999 REM U=MEAN
43000 X=RND(0)
43001 E=-U*LOG(1-X)
43002 RETURN
```

Program V.6: *Exponential random number generator subroutine (EXPONENT) using the RND function.*

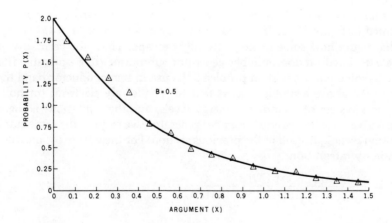

Figure V.11. *Statistical distribution for the exponential random number generator (EXPONENT), with some sample results for B = 0.5.*

V.7 The Fermi Distribution

Widely used in the study of particle physics, the Fermi distribution is based on the concept that no more than one indistinguishable entity can occupy a given cell (ie: an electron or hole in a crystal lattice) at a given time. The mathematical form of the associated probability density function is:

$$p(x) = \frac{1}{\ln(2)\,[1 + \exp(x)]} \qquad \text{for } x \geq 0 \qquad (V.22)$$

The shifted and scaled version is:

$$p(x) = \frac{\text{constant}}{1 + \exp[4(x - U)/V]} \qquad \text{for } x \geq 0 \qquad (V.23)$$

In this notation U represents the 50% point of the monotonically decreasing function, and V is a measure of the width of the transition region (see figure V.12).

As U becomes larger, the distribution as a whole becomes broader. As V becomes smaller, the transition region becomes sharper. For U = 1 and V small (V ≪ U), the Fermi distribution approximates the uniform distribution. For the case U < 0 and V ≪ ABS(U), it approximates the exponential.

Figure V.12 may be employed to visualize how the Fermi distribution can be used to deal with some common physical problems. Consider an undisturbed barrel filled with tennis balls. The vertical distribution of the balls could be treated in approximation as uniform up to the top of the pile, at which point the distribution function would abruptly go to zero. This would be the case U = 1 (the top of the undisturbed pile) and V = 0. If some agitation were introduced, the balls at the bottom of the pile would not move, but those at the top might jiggle up and down. No longer sharp, the transition region would have an appearance similar to that represented in figure V.12. If the lip of the barrel were at x = 1.4, it is apparent from that figure how some of the balls might escape. Thus, the Fermi distribution and the associated random number generator subroutine (to be presented) are applicable to some fairly common problems. Its use in semiconductor band theory is similar to the above example, except that the "balls" are electrons (or holes), the "barrel" is a system of quantized energy levels, and the "agitation" is thermal.

To derive the Fermi random number generator, we employ the same cumulative distribution technique used in the previous section. For the unshifted, unscaled form we obtain by integration:

$$Y = 1 - \frac{\ln[1 + \exp(X)]}{\ln(2)} + \frac{X}{\ln(2)} \qquad (V.24)$$

The objective is to express X in terms of Y as was done for the exponential distribution. However, a simple expression which achieves this end is not possible. Instead an iterative form can be written:

$$X = -\ln(2)(1 - Y) + \ln[1 + \exp(X)] \qquad (V.25)$$

Observe that X, the desired value, appears on both sides of the equation. By estimating X, putting that value into the right-hand side of the equation, and evaluating it, an improved estimate of X is obtained. This improved estimate can be used to get an even better estimate, and so on. The reader may satisfy himself that the above equation is at least qualitatively correct by observing that for Y = 0, a value X = 0 is obtained, and for Y = 1, a value X = infinity results.

In the shifted case the iteration equation is:

$$X = -(1 - Y)\ln[1 + \exp(U)] + \ln[\exp(U) + \exp(X)] \qquad (V.26)$$

The Fermi random number generator subroutine uses this iteration equation, along with some scaling, to create the desired results (see program V.7). The generator can be partially tested with a 1000-number sequence (see figure V.12).

Note that the Fermi distribution is continuous and has the property of being symmetrical about the 50% point. The iteration equation used is always convergent. The values of U and V are limited by the possibility of an overflow in the exponent function used in line 43027. For example, North Star BASIC (Release 3) requires that the absolute value of U/V be kept under 36.

Statements/Functions List

+, −, *, /, <
ABS, EXP, IF/THEN, GOTO, RND

Variables List

A, B, E, X, Y, Y1, U, V

Variables Passed to Subroutine

U, V

Table V.7 *Statements, functions, and variables used by the Fermi-distribution random number generator (FERMI).*

```
43023 REM FERMI RANDOM NUMBER GENERATOR (FERMI)
43024 REM U=INFLECTION POINT, V=SPREAD OF TRANSITION REGION
43025 X=RND(0)
43026 Y=1
43027 A=EXP(4*U/V)
```

Program V.7:

148 BASIC SCIENTIFIC SUBROUTINES

```
43028 B=(X-1)*LOG(1+A)
43029 Y1=B+LOG(A+EXP(Y))
43030 IF ABS((Y-Y1)/Y)<.001 THEN GOTO 43033
43031 Y=Y1
43032 GOTO 43029
43033 E=V*Y1/4
43034 RETURN
```

Program V.7: *Fermi distribution random number generator subroutine (FERMI) using the RND function.*

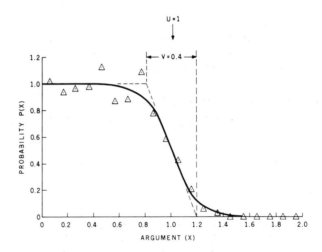

Figure V.12. *Statistical distribution for the Fermi random number generator (FERMI), with some sample results for $U = 1$; $V = 0.4$.*

V.8 The Cauchy Distribution

The form of the Cauchy distribution is:

$$p(x) = \frac{2U/\pi}{U^2 + x^2} \qquad \text{for } x \geq 0 \qquad (V.27)$$

This distribution exhibits the properties of an unbounded mean and standard deviation. It is useful in cases where the occurrence of large values is expected. One particular example is the diffusion of light in some papers. A point of light may be focused on a sheet of paper in one spot, but it emerges from the paper in a region surrounding that spot due to scattering. The distribution function best fitting the emerging light profile for some papers is the Cauchy.

The basic transformation from the uniform to the Cauchy distribution can be easily arrived at by using the cumulative distribution method. Performing the usual integration we get:

$$Y = (2/\pi) \arctan(X/U) \qquad (V.28)$$

Transforming this gives:

$$X = U \tan(\pi Y/2) \qquad (V.29)$$

The subroutine for providing Cauchy random number generation is shown in program V.8. Observe that because the tangent function is not available in many BASICs, sine/cosine is used.

A plot of the Cauchy distribution for $U = 1$ appears in figure V.13. The subroutine was used to create a 1000-number sequence to partially test the generator with the results also shown in that figure.

As was the case in previous sections, if the uniform distribution generator produces output values very near unity, overflows can result in the transformation, but that occurrence is highly unlikely.

Statements/Functions List

*

ATN (or ATAN), RND

Variables List

E, X, U

Variables Passed to Subroutine

U

Table V.8 *Statements, functions, and variables used by the Cauchy-distribution random number generator (CAUCHY).*

```
43048 REM CAUCHY RANDOM NUMBER GENERATOR (CAUCHY)
43049 REM U=MEAN
43050 X=RND(0)
43051 E=U*SIN(1.5707963267*X)/COS(1.5707963267*X)
43052 RETURN
```

Program V.8: *Cauchy random number generator subroutine (CAUCHY) using the RND function.*

150 BASIC SCIENTIFIC SUBROUTINES

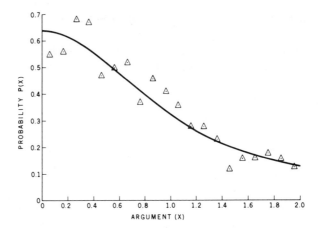

Figure V.13. *Statistical distribution for the Cauchy random number generator (CAUCHY), with some sample results for U = 1.*

V.9 The Gamma Distribution

In many statistical analyses, the random variable under consideration may show considerable skew in its distribution (see figure V.15). In such cases, the gamma (or Weibull; see section V.11) distribution may prove useful.

The basic form of the gamma distribution is (see reference 33):

$$p(x) = \frac{\exp(-x/B)\, x^{N-1}}{\Gamma(N)\, B^N} \qquad \text{for } x > 0; \; N, B > 0 \qquad (V.30)$$

Note that $\Gamma(N)$ is the gamma function (as distinguished from the gamma distribution). For $N = 1, 2, 3, 4, \ldots$, $\Gamma(N + 1) = N!$. The parameters for the gamma distribution are:

$$\text{mean} = NB \qquad (V.31)$$

$$\text{standard deviation} = B\sqrt{N}$$

In the case $N = 1$, the gamma distribution is equivalent to the exponential distribution. The case considered in this section is $N = 2$, which has the form:

$$p(x) = \frac{x \exp(-x/B)}{B} \qquad (V.32)$$

As the reader may see by comparing the figures in the sections on the exponential and Weibull distributions, the form corresponding to $N = 2$ offers a curve shape which falls between that of the exponential distribution and that of the $V = 2$ Weibull distribution.

By the cumulative distribution technique, the transformation between the uniform distribution random number Y and the gamma distribution variable X is:

$$z = -\ln[(1 - Y)/(1 + z)] \tag{V.33}$$
$$X = Bz$$

For simplicity and speed, the transformation is performed in two steps, one involving an iteration which is independent of the input parameter B and a second step which scales the output according to B. The iteration always converges. The subroutine employing these equations is shown in program V.9. Sample results are plotted in figure V.15.

There is a remote probability that an overflow will occur if the number generated by RND is sufficiently near unity. This is not guarded against since the probability of this happening is very low.

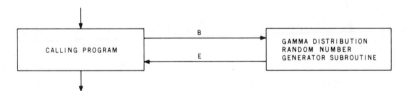

Figure V.14. *Subroutine connection for the gamma distribution random number generator (GAMMA).*

Statements/Functions List

$+, -, *, /, <$
ABS, IF/THEN, GOTO, LOG, RND

Variables List

B, E, X, Y, Y1

Variables Passed to Subroutine

B

Table V.9 *Statements, functions, and variables used by the gamma-distribution random number generator (GAMMA) for the case $N=2$. For $N=1$ see the exponential-distribution generator subroutine.*

```
43073 REM GAMMA (N=2) RANDOM NUMBER GENERATOR (GAMMA)
43074 REM B=INPUT PARAMETER
43075 Y=1
43076 X=RND(0)
43077 Y1=-LOG((1-X)/(1+Y))
43078 IF ABS((Y1-Y)/Y)<.001 THEN GOTO 43081
43079 Y=Y1
43080 GOTO 43077
43081 E=B*Y1
43082 RETURN
```

Program V.9: *Gamma distribution random number generator subroutine (GAMMA) using the RND function.*

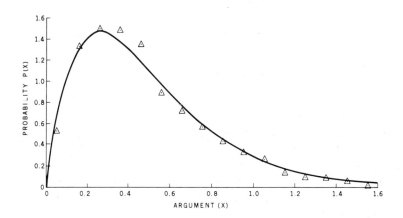

Figure V.15. *Statistical distribution for the gamma random number generator (GAMMA), with some sample results for B = 0.25.*

V.10 The Beta Distribution

This section considers a random number generator that has the property of being continuous, but which is defined only over the interval 0 to 1. By varying the input parameters, the distribution can take on many shapes. It is useful in Monte Carlo analyses in which the distributions of the input variables have been empirically determined and random number sequences following these distributions are desired.

The basic form of the beta distribution is:

$$p(x) = \frac{x^{A-1}(1-x)^{B-1}}{\Gamma(A)\Gamma(B)/\Gamma(A+B)} \qquad \text{for } 0 < x < 1; A, B > 0 \qquad (V.34)$$

Except for the cases $A < 1$ and $B < 1$, $p(x)$ is 0 at the endpoints 0 and 1. The case $A = 2$, $B = 3$ is shown in figure V.17. The parameters for the beta distribution are:

$$\begin{aligned} \text{mean} &= A/(A+B) \\ \text{standard deviation} &= \frac{1}{A+B}\left(\frac{AB}{A+B+1}\right)^{1/2} \end{aligned} \qquad (V.35)$$

For the example shown in figure V.17, the mean is 0.4 and the standard deviation is 0.2. For further information on the beta distribution, see references 26, 33 and 35.

The equations which transform the uniform distribution random numbers into those of the beta distribution may be found using the cumulative distribution integration technique. This was done for two cases, $A = 1$ and $A = 2$:

$$X = 1 - (1 - Y)^{1/B} \qquad \text{for } A = 1 \qquad (V.36)$$

$$X = 1 - \left(\frac{1-Y}{1+BX}\right)^{1/B} \qquad \text{for } A = 2$$

The first equation permits direct calculation of X; the second equation requires iteration. The beta distribution subroutine in program V.10 performs the required transformations.

The main precaution in using the subroutine relates to the power function (see lines 43107 and 43110). If B is small and Y is near unity, an overflow can result. The overflow would most likely occur in the logarithm and exponent subroutines which are usually employed to evaluate powers.

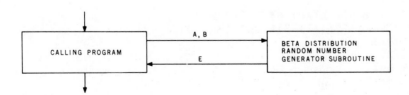

Figure V.16. *Subroutine connection for the beta distribution random number generator (BETA).*

Statements/Functions List

+, −, *, /, <, >,
ABS, GOTO, IF/THEN, RND

Variables List

A, B, E, X, Y, Y1

Variables Passed to Subroutine

A, B

Table V.10 *Statements, functions, and variables used by the beta-distribution random number generator subroutine (BETA).*

```
43096 REM BETA RANDOM NUMBER GENERATOR (BETA)
43097 REM INPUT PARAMETERS ARE A AND B
43098 REM A IS RESTRICTED TO A=1 AND A=2
43099 REM GUARD AGAINST DIVIDE BY ZERO
43100 IF B>0 THEN GOTO 43103
43101 E=1
43102 RETURN
43103 REM B>0
43104 IF A>2 THEN RETURN
43105 X=RND(0)
43106 IF A=2 THEN GOTO 43109
43107 E=1-(1-X)^(1/B)
43108 RETURN
43109 Y=1
43110 Y1=1-((1-X)/(1+B*Y))^(1/B)
43111 IF ABS((Y-Y1)/Y)<.001 THEN GOTO 43114
43112 Y=Y1
43113 GOTO 43110
43114 E=Y1
43115 RETURN
```

Program V.10: *Beta distribution random number generator (BETA) using the RND function.*

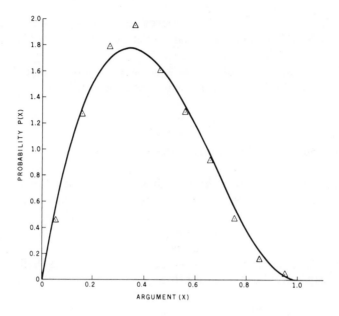

Figure V.17. *Statistical distribution for the beta random number generator (BETA), with some sample results for $A = 2$; $B = 3$.*

V.11 The Weibull Distribution

The Weibull distribution is very widely used in the industrial environment for the analysis of manufacturing tolerances and failures. The general form of the distribution is (see reference 13):

$$p(x) = \frac{Vx^{V-1}}{U^V} \exp[-(x/U)^V] \qquad \text{for } U, V, x > 0 \qquad (V.37)$$

The parameters of this distribution function are (see reference 33):

$$\text{mean} = U\Gamma(1 + 1/V) \qquad (V.38)$$
$$\text{standard deviation} = U\sqrt{\Gamma(1 + 2/V) - \Gamma^2(1 + 1/V)}$$

The shape of the Weibull distribution is largely affected by V, while the scale is controlled by U. Figure V.18 gives an indication of the influence of V. In failure analysis, the observed fitted values of V have the following physical interpretation:

V < 1   high initial failure rate which approaches 0 as x approaches infinity. It implies declining failure rate, characteristic of what is called *infant mortality*.

V = 1   the instantaneous failure rate is constant, implying that the failures are random with some characteristic time constant.

V < 1   distribution skewed to the left.

V = 3.2 distribution nearly Normal in shape.

V > 5   distribution skewed to the right. Implication is that there is a wear out or degradation involved, causing the failure rate to increase with time.

The case shown in figure V.19 corresponds to U = 0.5, V = 2.

The transformation equation can be easily derived by use of the cumulative distribution integral approach, since the analytic form of the Weibull cumulative probability distribution is (reference 13):

$$P(X) = 1 - \exp[-(x/u)^V] \qquad (V.39)$$

With Y as the random number derived from the uniform random number generator, we thus have:

$$X = U\{\ln[1/(1-Y)]^{1/V}\} \qquad (V.40)$$

The above transformation is employed in the Weibull random number generator subroutine shown in program V.11. The results of a 1000-number sequence are indicated in figure V.19.

Two potential problem areas in using the subroutine are in the logarithm and power functions. Values of the input random number near unity can cause an overflow in the logarithm function. Values near zero coupled with a small V can cause an overflow in the power function. These difficulties cannot be overcome by scaling because the curve shape is controlled by these parameters. Fortunately, the probability of an overflow is low.

In the next chapter, we continue consideration of approximations, focusing on functions such as the sine, cosine, logarithm and exponent.

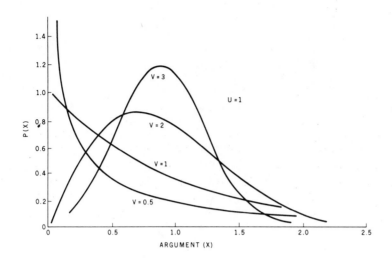

Figure V.18. *Various distributions for the Weibull random number generator.*

Statements/Functions List

$-, *, /,$
LOG, RND

Variables List

E, X, U, V

Variables Passed to Subroutine

U, V

Table V.11 *Statements, functions, and variables used by the Weibull-distribution random number generator subroutine (WEIBULL).*

```
43148 REM WEIBULL RANDOM NUMBER GENERATOR (WEIBULL)
43149 REM INPUT PARAMETERS ARE U AND V
43150 X=RND(0)
43151 E=U*((LOG(1/(1-X)))^(1/V))
43152 RETURN
```

Program V.11: *Weibull distribution random number generator subroutine (WEIBULL) using the RND function.*

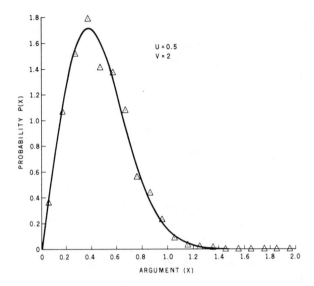

Figure V.19. *Statistical distribution for the Weibull random number generator (WEIBULL), with some sample results for U = 0.5; V = 2.*

CHAPTER VI

BASIC SERIES APPROXIMATIONS

Any reasonably well-behaved function can be "expanded" into what is generally called a Taylor (polynomial) series. However, such expansions are seldom the best for computer evaluation, especially when only a limited range of the independent variable is under consideration. Under the criterion of minimizing the maximum error in the approximation interval, the *minimax* or *optimal* series expansion is better than the corresponding Taylor series. Section VI.1 briefly describes the Taylor series expansions for sin(x) and arctan(x) and focuses on convergence and accuracy. The following section presents the concept of the minimax polynomial and demonstrates why this expansion is superior to the Taylor series for computer approximation. The discussion continues in section VI.3, showing that by a change of variables, the power of the general optimal series expansion can be employed to approximate some very common and important functions. Tables of series coefficients are included.

Section VI.4 contains the core of the chapter. A simple, common, and very general approximation technique, least-squares fitting to function tables, is presented as a ready compromise to minimax polynomial generation. High accuracy approximations are given, and the coefficient tables are incorporated into series-approximation subroutines. The functions that are approximated and the argument ranges covered, as well as the resulting accuracies, are given in this section.

The discussion in this chapter is longer than usual. The intent is to provide not only specific function-approximation subroutines, but also some guidelines which will enable the reader to generate further routines as the need arises.

VI.1 Taylor Series Expansions

Generally, any real, continuous function f(x) having defined derivatives can be expressed as a polynomial expansion about a reference point, x_0, with the long hand representation:

$$f(x) = f(x_0) + f^{(1)}(x_0)(x - x_0) + f^{(2)}(x_0)(x - x_0)^2/2! + \cdots \qquad (VI.1a)$$

Observe that $f^{(n)}(x_0)$ is defined as the nth order derivative of $f(x)$ evaluated at $x = x_0$. The shorthand equivalent of equation (VI.1a) is:

$$f(x) = \sum_{n=0}^{\infty} f^{(n)}(x_0) \frac{(x - x_0)^n}{n!}$$

or

$$f(x) = \lim_{n \to \infty} f_n(x) \qquad (VI.1b)$$

where

$$0! \equiv 1$$
$$f^{(0)}(x_0) \equiv f(x_0)$$
$$f_n(x) \equiv \sum_{m=0}^{n} f^{(m)}(x_0) \frac{(x - x_0)^m}{m!}$$

Equation (VI.1b) is in a form which is convenient for loop evaluation, which may account for its popularity with computer programmers.

Many programmers who are faced with approximating a function immediately apply a Taylor series expansion along with a convergence test, such as:

$$|f_n(x) - f_{n-1}(x)| < \epsilon \qquad (VI.2)$$

Presumably, when the difference between the approximation using n terms and that using $n - 1$ terms is less than ϵ, the accuracy of $f_n(x)$ in approximating $f(x)$ is better than ϵ. This assumption can be grossly in error depending upon the function being evaluated. Slowly converging series often present problems in this respect.

The reasons for not indiscriminantly using equations (VI.1) or (VI.2) may be demonstrated by considering two common functions, $\sin(x)$ and $\arctan(x)$.

The Taylor series expansion around $x_0 = 0$ (MacLaurin series) for $\sin(x)$ is:

$$\sin(x) = x - \frac{x^3}{3!} + \frac{x^5}{5!} - \cdots \qquad (VI.3a)$$

or

$$\sin(x) = \sum_{n=0}^{\infty} (-1)^n \frac{x^{(2n+1)}}{(2n + 1)!} \qquad (VI.3b)$$

If we apply equation (VI.2) as the test for accuracy, we have:

$$\frac{x^{(2n+1)}}{(2n+1)!} < \epsilon \tag{VI.4}$$

If x is 1 radian (57°) and the required accuracy is 10^{-8}, then equation (VI.4) indicates that seven terms are required in the series for the accuracy specified. The test for *residual* may be done, in principle, more correctly by noting that the series is uniformly convergent, having terms alternating in sign. Thus, the absolute error of the approximation is less than the absolute value of the last term included. Therefore we again conclude that seven terms are needed for 10^{-8} accuracy in approximately calculating sin(57°). In the case of alternating series having terms which monotonically decrease in absolute value, equation (VI.2) is applicable.

So far we have assumed the computer to be perfect in terms of round-off error. In some interpreters and compilers having eight-place accuracy, the computer rounds off to the 8th decimal place by consistently rounding down or up. Quite often the round-off direction is down, as this corresponds to simple truncation. In those cases, the ensuing error in calculating n terms in a series expansion is on the order of $\pm(n/2) \times 10^{-8}$*. For the sin(x) expansion example given above, the expected truncated series accuracy of 10^{-8} would be reduced to approximately 10^{-7} because of round-off error. The next section shows that a different series expansion for sin(x) can be used, containing only five terms and giving an error of less than 10^{-8} before round-off, along with a more accurate answer after round-off.

These conclusions are not very exciting largely because the Taylor series expansion for sin(x) in the first quadrant is very rapidly convergent and thus quite adequate. Another important function, arctan(x), is not as quickly convergent in some regions of its *limited* convergence interval. The MacLaurin series expansion for arctan(x) is:

$$\arctan(x) = x - \frac{x^3}{3} + \frac{x^5}{5} - \frac{x^7}{7} + \cdots$$

$$= \sum_{n=0}^{\infty} (-1)^n \frac{x^{(2n+1)}}{2n+1} \quad \text{for } -1 < x \leq 1 \tag{VI.5}$$

This expansion has difficulty converging near $x = \pm 1$. At $x = -1$ the series diverges, although a finite answer exists ($\pi/4$), whereas at $x = 1$ the series converges very slowly to $\pi/4$, or 45° [tan(45°) = 1; arctan(1) = $\pi/4$ (= 45°)]. Only the region between 0° and 45° needs to be used for the expansion, as trignometric identities exist for extending the inversion to other regions (eg: arctan(x) = $\pi/2$ − arctan(1/x) for x > 1). The technique of *range reduction* for improving accuracy will be considered in a later section.

---

*In "smarter" software, rounding is done to the *closest* level, leading to a final error which is proportional to $\sqrt{n} \times 10^{-8}$.

Consider the use of equation (VI.5) when x is near 1. Using equation (VI.2) to determine the number of terms required for 10^{-8} accuracy, we get n = 5×10^7. Applying the alternating sign convergence test also leads to N = 5×10^7. In either case, this series expansion for arctan(x) is unusable. Even if the computer were fast enough to calculate the approximation in an acceptable length of time, the ensuing round-off error would be prohibitive. The next section shows the existence of minimax series expansions accurate to better than (in terms of relative error) 4×10^{-8} which have only eight terms. The series presented are operable either over $1 \le |x|$ (note the equality sign) or over $0 \le x < \infty$.

We observe that the error test [equation (VI.2)] would fail dismally for the Taylor series expansion of arctan(x) near x = -1. It would again predict that on the order of 5×10^{-7} terms would be required for 10^{-8} accuracy, while in actuality the series diverged. In this case, the Taylor series expansion is not rapidly converging, thus leading to a failure in equation (VI.2).

In general, equation (VI.2) is practical if:

1) the series has decreasing alternating sign terms

   or

2) the series has decreasing terms in which the absolute value of the ratio of neighboring terms is greater than 10 for all pairs of terms past the termination point. That is, if:

$$f(x) = \sum_{n=0}^{\infty} c_n x^n$$

then beyond the termination point it is required that:

$$\left| \frac{c_n x^n}{c_{n+1} x^{n+1}} \right| = \left| \frac{c_n}{c_{n+1} x} \right| > 10$$

Observe that this latter restriction is only valid for MacLaurin series expansions. For Taylor series expansions, x is replaced with x − x_0. Note that, although small values of (x − x_0) are conductive to rapid convergence, they are not a sufficient condition for convergence.

VI.2 Approximate Series Expansions

Taylor series expansions have many nice properties. However, one of the less desirable features is that they are *not* the optimal expansions for a given argument interval when a truncated series is to be used. A simple example will clarify this.

Consider the single term approximation to $\sin(x)$ over the interval $0 \leq x \leq \pi/2$ (see figures VI.1 and VI.2). Although the accuracy of $\sin(x) \approx x$ is good when x is small, it leads to a 0.57 error at $\pi/2$. If we instead wish to minimize the maximum absolute error over that interval, $\sin(x) \approx 0.73x$ is a much better approximation. The maximum absolute error is less than 0.15 in this case.

If our criteria were instead to minimize the relative error (eg: percent deviation), the MacLaurin series single-term truncation leads to 57% error at $\pi/2$, while the previous minimized (relative to maximum absolute error) approximation is off by less than 27%, with the maximum relative error occurring at $z = 0$ (see figure VI.2). The maximum relative error can be further reduced to about 22% by the approximation $\sin(x) \approx 0.78x$. The maximum relative errors in this case occur at $x = 0$ and $x = \pi/2$. This series is considered to be the *minimax* or optimal series expansion for $\sin(x)$ over $-\pi/2 \leq x \leq \pi/2$, given that one term is allowed.

The above example illustrates the fact that either by the criterion of absolute error or by the criterion of relative error, the truncated MacLaurin series for $\sin(x)$ is not optimal for approximation. Also observe that the coefficient (0.73 for absolute error or 0.78 for relative error) in the optimal approximation is dependent on the interval chosen. The more the interval is restricted to the region surrounding $x = 0$, the closer the coefficient will be to unity, and the better the fit. In general, the more the interval is restricted to that immediately surrounding the Taylor series expansion point, the more the coefficients in the optimal series approximate those of the Taylor series expansion. This is because the highest convergence rate occurs near the expansion point, and it is difficult to do better than that.

Continuing with the $\sin(x)$ example, table VI.1 shows the series coefficients which would be used in the MacLaurin and optimal series expansions to give a desired accuracy (excluding round-off) of approximately 10^{-8}. Observe that because the relative error is the criterion, the first coefficient in the optimal series is the same (to eight places) as that in the MacLaurin series. This is required because for very small x, $\sin(x) \approx x$. Observe, however, that all the other coefficients are different.

A corresponding series-expansion comparison for $\arctan(x)$ appears in table VI.2. Note that about 10^7 terms are required in the MacLaurin series at $x = 1$ to give the same order of accuracy as the optimal series evaluated at that point. The situation for the MacLaurin series is even worse in the vicinity of $x = -1$.

Also interesting to note is that in the series expansion comparison in table VI.2, the signs of the coefficients are conserved, and the coefficients themselves bear some resemblance to one another up to the fifth term. After that, the coefficients look quite different.

164 BASIC SCIENTIFIC SUBROUTINES

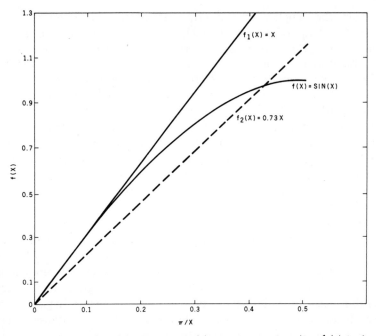

Figure VI.1. *Single term approximations to $\sin(x)$ over $0 \leq x \leq \pi/2,$: $f_1(x)$ is the truncated MacLaurin series expansion; $f_2(x)$ is the single term fit minimizing the maximum error.*

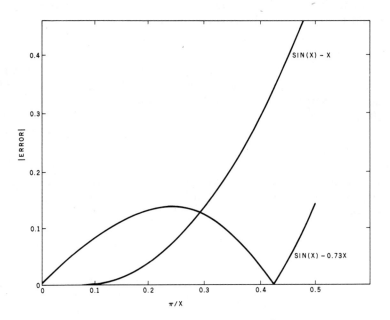

Figure VI.2. *Errors due to single-term approximations to $\sin(x)$.*

Term	MacLaurin Coefficients	Optimal Coefficients
X	+ 1.00000000	+ 1.000000000
X^3	− 0.16666667	− 0.1666665669
X^5	+ 0.00083333333	+ 0.000833302517
X^7	− 0.00019841270	− 0.000198074143
X^9	+ 2.755732 × 10^{-6}	+ 2.601886908
X^{11}	− 2.5052109 × 10^{-8}	Series Truncated
X^{13}	+ 1.6059045 × 10^{-10}	Series Truncated

Table VI.1 *Coefficients for the MacLaurin and optimal series expansions for sin(x) (see also references 3 and 37). The interval optimized over is $-\pi/2 \leq x \leq \pi/2$. In this interval the optimal (relative error) series is accurate to better than about 5×10^{-9}.*

Term	MacLaurin Coefficients	Optimal Coefficients
X	+ 1.00000000	+ 0.99999933
X^3	− 0.33333333	− 0.33329856
X^5	+ 0.20000000	+ 0.19946536
X^7	− 0.1428571	− 0.13908534
X^9	+ 0.11111111	+ 0.096420044
X^{11}	− 0.090909090	− 0.055909886
X^{13}	+ 6.07692308	+ 0.021861229
X^{15}	− 0.066666667	− 0.0040540580

Table VI.2 *Coefficients for the MacLaurin and optimal series expansions for arctan(x) (see also references 3 and 37). The relative accuracy is better than 4×10^{-8}. The interval of convergence for the MacLaurin series is $-1 < x \leq 1$, and for the optimal series it is $-1 \leq x \leq 1$.*

VI.3 Variations on the Optimal Series Theme

The optimal series coefficients given in table VI.2 for the arctan(x) approximation were stipulated to be optimal over the interval $-1 \leq x \leq 1$. This restricts us to the angular range of $\pm 45°$ about $0°$. A trigonometric identity was given earlier which allowed continuation to $45-90°$. However, this extra range-changing step is not necessary since another series expansion exists which is optimized over $0 \leq x \leq \infty$. However, it does not have a simple x^n dependence:

$$\arctan(x) = \frac{\pi}{4} + \sum_{n=0}^{7} C_n \left(\frac{x-1}{x+1}\right)^{(2n+1)} \qquad (VI.6)$$

This series uses the same coefficients as given in table VI.2 and also gives the same order of relative accuracy. To execute this series on a computer, simply make the transformation $x \leftarrow (x-1)/(x+1)$ and proceed as before.

Hasting's book *Approximations for Digital Computers* (reference 37) presents several such optimal expansions, a few of which are presented in tables VI.3 through VI.5. Some selection/alteration has been made in order to put this information into a form readily usable with typical microcomputer software.

The optimal series approximations given in tables VI.1 through VI.5 are not implemented in the subroutines given in this chapter. Instead, the coefficients used are derived from least-squares series regressions which approximate the properties of the optimal series. This is discussed in the next section.

Form: $\qquad \log_{10} x = \frac{1}{2} + \sum_{n=0}^{4} C_n \gamma^{(2n+1)}$

where $\gamma = (x - \sqrt{10})/(x + \sqrt{10})$

Coefficients:
$C_0 = 0.86859172$
$C_1 = 0.28933552$
$C_2 = 0.17752207$
$C_3 = 0.094376476$
$C_4 = 0.19133771$

Table VI.3 *Minimax polynomial expansion for* $\log_{10} x$. *The applicable range is* $1 \le x \le 10$. *Relative error is better than* 1.5×10^{-7} *(see reference 3). Note that* $\log_e X = (\log_{10} X)/(\log_{10} e)$.

Form: $\qquad 10^x = 1 + \sum_{n=1}^{7} C_n x^n$

Coefficients:
$C_1 = 1.15128759$
$C_2 = 0.66284315$
$C_3 = 0.25360332$
$C_4 = 0.075467547$
$C_5 = 0.013420940$
$C_6 = 0.005654902$

Table VI.4 *Minimax polynomial expansion for* 10^x. *The applicable range is* $0 \le x \le 1$. *Relative error is better than* 1.5×10^{-7} *(see reference 37).*

Form: $$e^{-x} = \left[\sum_{n=0}^{6} C_n x^n \right]^{(-4)}$$

Coefficients:
$C_0 = 1.00000000$
$C_1 = 0.24999868$
$C_2 = 0.031257583$
$C_3 = 0.0025913712$
$C_4 = 0.0001715620$
$C_5 = 0.0000054302$
$C_6 = 0.0000006906$

Table VI.5 *Minimax polynomial expansion for e^{-x}. The applicable range is $0 \le x < \infty$. Relative accuracy is better than 3×10^{-7} (see references 3 and 37).*

VI.4 Least-Squares Regression

Over a limited argument interval and with a limited number of terms, the Taylor series approximation is not usually the best one for computer evaluation; the optimal series is better. These series may be obtained by techniques called *economization* or *telescoping* (see references 9, 37, and 39). One straightforward approach to obtaining near-minimax approximations is to use Chebyshev polynomial expansions. In practice, the interval of approximation is reduced to $(-1, +1)$, and the function is expanded into a limited Chebyshev polynomial series. The corresponding power series can then be derived, and the range transformed back to the original limits. The problem is that the Chebyshev polynomial expansions are not common tabulations. However, an alternate technique is to expand the function into a standard Taylor series, find the corresponding Chebyshev series term-by-term, and then perform the truncation with respect to the Chebyshev series. This is discussed in detail in Volume II. In this chapter, an alternate method is proposed: least-squares approximation.

The basic concept in minimax approximations is to reduce all the excursions from the true function to a minimum level consistent with the number of terms used in the fitting. In the true minimax polynomial, this is achieved rigorously. The maximum excursion obtained is the minimum possible. If the same number of terms were used, but the fitting were via a least-squares criterion, the fit would be biased toward the minimization of the larger differences at the expense of the smaller differences. That is, whereas the minimax approximation curve may wander around the true function with several extreme, but all equal, differences of E, the least-squares variation extreme differences would be E1, E2, E3, and so on, with some larger and some smaller than E. There is still a maximum, however. Thus, least-squares fitting can be applied as a usable approximation to the minimax approximation. However, more terms may be required to insure that the maximum error is less than some given bound.

The value in using the least-squares criterion is the ease in which the mechanics of the approximation can be performed by using a regression package with double-precision arithmetic. For example, high-accuracy polynomial expansions can be generated by inputting function values from standard math tables. This approach was used to generate the approximations referred to in table VI.6.

The mechanics for generating these approximations were simple. Reference 30 was used for the high-accuracy function tables. The least-squares polynomial approximation routine given in Volume II was employed for the regression, although the regression program in reference 17 could have also been used (if the low-accuracy exponentiation function were replaced with simple multiplications). The least-squares regression was performed using double-precision MITS BASIC. The polynomial-expansion coefficients obtained are shown in tables VI.8 through VI.14.

As evidenced by tables VI.6, VI.8 and VI.9, the sine and cosine series can be approximated to high accuracy with a relatively short and simple power series. The arctangent approximation requires more terms to achieve lower accuracy, which is the result of the slow convergence discussed in section VI.1. Note that the range of the arctangent approximation can be altered with a change of variables $Z = (X - 1)/(X + 1)$. The regression was performed over the interval $0 \leq X \leq 1$, but the results can be used for an approximation over the interval $0 \leq X < \infty$.

The two logarithm functions were regressed with respect to the variable $Z = (X - C)/(X + C)$. This variable choice was obtained by consulting the lower accuracy minimax approximations given in reference 37. This points to an important rule. Before embarking on an approximation, check to see if there is a transformation of variables which will aid in convergence. This will be discussed more fully in the last section.

The exponent functions are largely simple polynomial expansions in X. However, the power-of-ten approximation points to another consideration involving accuracy. Observe that the series obtained is squared; the variable actually regressed against was $X/2$. This served to reduce the range of variation in the function and permitted a better fit. The range was recovered by simply squaring the result.

The development of the subroutines which apply the coefficients shown in the tables required consideration of both range and accuracy. These subroutines are shown in programs VI.1 through VI.8, and data relevant to these subroutines are given in tables VI.7 and VI.15 through VI.19. The routines, though written in BASIC, are generally only a factor of four to five times slower than the subroutines provided internal to the interpreter, and the accuracy is better (as will be seen later). The execution times are not more divergent because both the internal machine language and BASIC subroutines call the same floating-point math routines. In arithmetic-intensive operations, the fact that the BASIC statement must be interpreted becomes less important.

The general input/output structure of the subroutines appears in figure VI.3. The variable X is transferred to the main subroutine, which directs further program action. The range of the argument is reduced to that acceptable to the approximation (see table VI.6). A variable transformation is performed if needed. The series expansion coefficients are called from a data subroutine, and the summation is performed by a call to another subroutine (SERSUM). The result is adjusted if necessary and returned to the calling program. A separate coefficient subroutine (<name>DATA, see Appendix Ib) is used for modularity. The accuracy of the approximation can be altered by simply changing the coefficient values, A(I), and the number of coefficients, N + 1, in just that section of the program. A separate summation subroutine, SERSUM, which is used repeatedly, is provided to save space. The summation subroutine uses a nested evaluation technique (see reference 22) to reduce the number of multiplications required. Also, it provides some advantages with regard to round-off error, especially in series having coefficients which decrease with the coefficient index. Note that this is not the case with the logarithm approximations.

The coefficient and summation subroutines are simple, requiring no further explanation. The main subroutines are complicated largely by the range reduction. In the case of the sine and cosine, the range must be reduced to the first quadrant. This is done by first reducing the angle range to $0 \leq X < \infty$ and then to $0 \leq X < 2\pi$. A series of tests is then made to determine which quadrant the angle falls in, and the argument is adjusted accordingly.

For the arctangent subroutine, the range is first reduced to $0 < X < \infty$. At this point one might simply apply the series approximation as indicated by the equation at the bottom of table VI.10. However, there are some problems with round-off error which can be greatly (a thousandfold!) diminished by using the variable X for $X < 0.5$ and the variable $(X - 1)(X + 1)$ otherwise (this was empirically determined).

Range reduction in the logarithm subroutines consists largely of inverting the argument if it is less than unity and repeatedly dividing by 10 (or e, depending on whether the function to be approximated is the base 10 logarithm or natural logarithm, respectively), until the desired range is achieved. Note that the repeated division is a potential source of round-off error. The number of divisions is called the *characteristic*, and the series calculation gives the *mantissa*. The power-of-ten (or e) subroutines work in reverse. The characteristic is determined from the integer portion of the argument, and the mantissa from the remainder. The series (10^x and $\exp(x)$) use the mantissa, and the result is repeatedly multipled according to the characteristic.

The accuracies given in table VI.6 refer to the situation in which there is no round-off error. Examples of real situations are shown in listings VI.1 through VI.4. The demonstration programs used to generate these listings were programs VI.9 through VI.12.

The demonstration programs were run in North Star BASIC, both eight- and fourteen-digit precision. In general, one or two digits of accuracy are lost in the approximation due to round-off. This may be seen from the eight-digit exercises where the approximations are known to be better than eight digits. For the fourteen-digit examples, the results are in general agreement with table VI.6. Note, however, that the comparison for accuracy is with the function supplied by North Star BASIC. There are several cases in which the approximation is better than the function subroutine internal to BASIC. For example, in the fourteen-digit arctangent approximation, the estimate is accurate to the full fourteen digits for arctan(1), but the North Star function (ATN) is in error in the twelfth position. This points out a problem in testing such high-accuracy subroutines. They can be more precise than already packaged routines, and we must resort to the math tables!

In summary, the major concern in using the subroutines in this chapter is round-off error. Generally, if two more digits of computing precision are available beyond the digit accuracy indicated in table VI.6, there should be little difficulty. In terms of the allowable inputs, the subroutines all reduce the range of the argument so that there is considerable freedom, except for the logarithm subroutine. The logarithm of a negative number is not defined, and the subroutines do not test for this. In most cases, the only effect will be to return a nonsense result. However, there is also the possibility of an overflow.

Overflow is also a potential problem in the power subroutines. If the argument is too large, the resulting number will be outside the range of the software. Most software containing the log and power functions have a similar limitation.

Series	No. Terms	Range	Accuracy
sin	7	$-\pi/2, \pi/2$	10^{-14}
cos	7	$-\pi/2, \pi/2$	10^{-12}
exp	9	0, 1	5×10^{-10}
$\log_e$	10	1, e	2×10^{-11}
arctan	11	-1, 1	3×10^{-10}
arctan	12	$0, \infty$	3×10^{-10}
$\log_{10}$	11	1, 10	2×10^{-10}
10^x	10	0, 1	8×10^{-10}

Table VI.6 *Summary of the least-squares series expansion specifications.*

Subroutine	Size (bytes)	BASIC Function (seconds)	North Star/8 (seconds)	North Star/14 (seconds)
Series summation	96	.....	.....	.....*
Sine	500	0.1	0.32	0.46
Cosine	486	0.1	0.34	0.48
Arctangent	569	0.13	0.40	0.54
Log	560	NA	0.44	0.61
Ln	554	0.1	0.54	0.77
10^x	541	NA	0.38	0.47
Exp(x)	518	0.07	0.37	0.46

*Depends on the series evaluated

Table VI.7 *Size and average execution times of the least-squares series approximation subroutines presented in this section. "BASIC Function" refers to the internal subroutine provided with the eight-digit North Star interpreter (Version 6, Release 4). "North Star/8" refers to the BASIC subroutine execution time when run in eight-digit precision on North Star BASIC Version 6, Release 2. "North Star/14" similarly refers to the fourteen-digit precision case. The fourteen-digit evaluation takes approximately 40% more time than the eight-digit evaluation.*

$$SIN(X) = X + A_1X^3 + A_2X^5 + A_3X^7 + A_4X^9 + A_5X^{11} + A_6X^{13}$$

Coefficient	Value
A_1	−0.1666666666671334
A_2	0.00833333333809067
A_3	−0.000198412715551283
A_4	0.0000027557589750762
A_5	−0.00000002507059876207
A_6	0.000000000164105986683

Table VI.8 *Series approximation for sin(X) in the interval $-\pi/2 \le X \le \pi/2$. The accuracy of the approximation is roughly 10^{-14}.*

$$\text{COS}(X) = X + A_1 X^2 + A_2 X^4 + A_3 X^6 + A_4 X^8 + A_5 X^{10} + A_6 X^{12}$$

Coefficient	Value
A_1	−0.4999999999982
A_2	0.04166666664651
A_3	−0.001388888805755
A_4	0.000024801428034
A_5	−0.0000002754213324
A_6	0.0000000020189405

Table VI.9 Series approximation for cos(X) over the interval $-\pi/2 \le X \le \pi/2$. The accuracy of the approximation is roughly 10^{-12}.

$$\text{(a) ARCTAN}(X) = X + A_1 X^3 + A_2 X^5 + A_3 X^7 + A_4 X^9 + A_5 X^{11} + A_6 X^{13} + A_7 X^{15} + A_8 X^{17} + A_9 X^{19} + A_{10} X^{21}$$

Coefficient	Value
A_1	−0.333333311286
A_2	0.199998774421
A_3	−0.142831560376
A_4	0.110840091104
A_5	−0.089229124381
A_6	0.070315200033
A_7	−0.049278908030
A_8	0.026879941561
A_9	−0.009568384520
A_{10}	0.001605444922

$$\text{(b) ARCTAN}(X) = 0.7853981634 + Z + A_1 Z^3 + A_2 Z^5 + A_3 Z^7 + A_4 Z^9 + A_5 Z^{11} + A_6 Z^{13} + A_7 Z^{15} + A_8 Z^{17} + A_9 Z^{19} + A_{10} Z^{21}$$

where $Z = (X-1)/(X+1)$

Table VI.10 Table VI.10a shows the series approximation for arctan(x) over the interval $-1 \le X \le 1$. The accuracy of the approximation is roughly 3×10^{-10}. The same coefficients used in the series shown in table VI.10b give the same order of approximation over the range $0 \le X < \infty$.

$$LOG_{10}(X) = 0.5 + A_1Z + A_2Z^3 + A_3Z^5 + A_4Z^7$$
$$+ A_5Z^9 + A_6Z^{11} + A_7Z^{13} + A_8Z^{15}$$
$$+ A_9Z^{17} + A_{10}Z^{19}$$

where $Z = (X-c)/(X+c)$ $c = 3.162277660$

Coefficient	Value
A_1	0.8685889644
A_2	0.2895297117
A_3	0.1737120608
A_4	0.1242584742
A_5	0.0939080460
A_6	0.1009301264
A_7	−0.0439630355
A_8	0.3920576195
A_9	−0.5170494708
A_{10}	0.4915571108

Table VI.11 *Series approximation for $log_{10}(X)$ over the interval $1 \leq X \leq 1$. The accuracy of the approximation is roughly 2×10^{-10}.*

$$LOG_e(X) = 0.5 + A_0Z + A_1Z^3 + A_2Z^5 + A_3Z^7 + A_4Z^9$$
$$+ A_5Z^{11} + A_6Z^{13} + A_7Z^{15} + A_8Z^{17} + A_9Z^{19}$$

where $Z = (X-c)/(X+c)$ $c = 1.6487212707$

Coefficient	Value
A_0	2.00000000000
A_1	0.66666672443
A_2	0.3999895288
A_3	0.286436047
A_4	0.197959107
A_5	0.6283533
A_6	−4.54692
A_7	28.117
A_8	−86.42
A_9	106.1

Table VI.12 *Series approximation for $log_e(X)$ over the interval $1 \leq X \leq e$. The accuracy of the approximation is roughly 2×10^{-11}.*

$$10^X = (1 + A_1X + A_2X^2 + A_3X^3 + A_4X^4 + A_5X^5 + A_6X^6 + A_7X^7 + A_8X^8 + A_9X^9)^2$$

Coefficient	Value
A_1	1.1512925485
A_2	0.6627373050
A_3	0.2543345675
A_4	0.0732032741
A_5	0.0168603036
A_6	0.0032196227
A_7	0.0005547662
A_8	0.0000573305
A_9	0.0000179419

Table VI.13 Series approximation for 10^X over the range $0 \leq X \leq 1$. The accuracy of the approximation is roughly 8×10^{-10}.

$$e^X = 1 + A_1X + A_2X^2 + A_3X^3 + A_4X^4 + A_5X^5 + A_6X^6 + A_7X^7 + A_8X^8$$

Coefficient	Value
A_1	0.99999999668
A_2	0.49999995173
A_3	0.16666704243
A_4	0.04166685027
A_5	0.00832672635
A_6	0.00140836136
A_7	0.00017358267
A_8	0.00003931683

Table VI.14 Series approximation for e^X over the interval $0 \leq X \leq 1$. The accuracy of the approximation is roughly 5×10^{-10}.

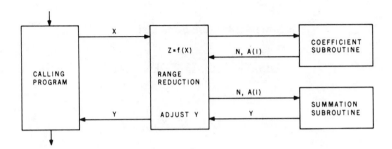

Figure VI.3. *Subroutine connection for the least-squares series approximations given in Chapter VI.*

Statements/Functions List

$+, -, *,$
FOR/NEXT/STEP

Variables List

$A(I), I, N, Y, Z$

Variables Passed to Subroutine

$A(I), N$

Table VI.15 *Statements, functions, and variables used in the basic series-summation subroutine (SERSUM).*

```
43199 REM SERIES SUMMATION SUBROUTINE (SERSUM)
43200 Y=0
43201 FOR I=N TO 0 STEP -1
43202 Y=Z*Y+A(I)
43203 NEXT I
43204 RETURN
```

Program VI.1: *General series summation subroutine (SERSUM) called by other subroutines. Nested evaluation is used. See reference 22.*

Statements/Functions List

−, *, /, <, >
ABS, GOSUB, IF/THEN, INT

Variables List

A(I), N, X, X1, X2, X3, Y, Z

Variables Passed to Subroutine

X

Table VI.16 *Statements, functions, and variables used by the sine and cosine approximation subroutines (SINE, COSINE). The items in table VI.15 should be added to this list.*

(a)
```
43209 REM SINE SERIES SUBROUTINE (SINE)
43210 X1=1
43211 IF X<0 THEN X1=-1
43212 X3=ABS(X)
43213 X2=3.141592653589793
43214 REM REDUCE RANGE
43215 X3=X3-2*X2*INT(.5*X3/X2)
43216 IF X3>X2 THEN X1=-X1
43217 IF X3>X2 THEN X3=X3-X2
43218 IF X3>X2/2 THEN X3=X2-X3
43219 Z=X3*X3
43220 GOSUB 43300
43221 GOSUB 43200
43222 Y=X1*X3*Y
43223 RETURN
```

(b)
```
43299 REM SINE SERIES COEFFICIENTS (SINDATA)
43300 N=6
43301 A(0)=1
43302 A(1)=-.1666666666671334
43303 A(2)=.00833333333809067
43304 A(3)=-.000198412715551283
43305 A(4)=.0000027557589750762
43306 A(5)=-.00000002507059876207
43307 A(6)=.0000000000164105986683
43308 RETURN
```

Program VI.2: *Sine series-approximation subroutine 2a (SINE) with its associated coefficients (2b). Also uses program VI.1.*

```
43224 REM COSINE SERIES SUBROUTINE (COSINE)                (a)
43225 X2=3.141592653589793
43226 X1=1
43227 REM REDUCE RANGE
43228 X3=ABS(X)
43229 X3=X3-2*X2*INT(0.5*X3/X2)
43230 IF X3>X2 THEN X1=-1
43231 IF X3>X2 THEN X3=X3-X2
43232 IF X3>X2/2 THEN X1=-X1
43233 IF X3>X2/2 THEN X3=X2-X3
43234 Z=X3*X3
43235 GOSUB 43310
43236 GOSUB 43200
43237 Y=X1*Y
43238 RETURN


43309 REM COSINE SERIES COEFFICIENTS (COSDATA)             (b)
43310 N=6
43311 A(0)=1
43312 A(1)=-.4999999999982
43313 A(2)=.04166666664651
43314 A(3)=-.001388888805755
43315 A(4)=.000024801428034
43316 A(5)=-.0000002754213324
43317 A(6)=.0000000020189405
43318 RETURN
```

Program VI.3: *Cosine series-approximation subroutine 3a (COSINE) with its associated coefficients (3b). Also uses program VI.1.*

Statements/Functions List

$+, -, *, <$
ABS, GOSUB

Variables List

A(I), N, X, X1, X3, Y, Z, Z1

Variables Passed to Subroutine

X

Table VI.17 *Statements, functions, and variables used by the arctangent subroutine (ARCTAN). The items shown in table VI.15 should be included in the list below.*

(a)
```
43244 REM ARCTANGENT SERIES SUBROUTINE (ARCTAN)
43245 X1=1
43246 IF X<0 THEN X1=-1
43247 X3=ABS(X)
43248 Z1=(X3-1)/(X3+1)
43249 IF X3<.5 THEN Z1=X3
43250 Z=Z1*Z1
43251 REM GET SERIES COEFFICIENTS
43252 GOSUB 43320
43253 REM SUM SERIES
43254 GOSUB 43200
43255 Y=Z1*Y
43256 IF X3>=.5 THEN Y=Y+.78539816339745
43257 Y=X1*Y
43258 RETURN
```

(b)
```
43319 REM ARCTANGENT SERIES COEFFICIENTS (ARCDATA)
43320 N=10
43321 A(0)=1
43322 A(1)=-.333333311286
43323 A(2)=.199998774421
43324 A(3)=-.142831560376
43325 A(4)=.110840091104
43326 A(5)=-.089229124381
43327 A(6)=.070315200033
43328 A(7)=-.049278908030
43329 A(8)=.026879941561
43330 A(9)=-.009568384520
43331 A(10)=.001605444922
43332 RETURN
```

Program VI.4: Arctangent series-approximation subroutine 4a (ARCTAN) with its associated coefficients (4b). Also uses program VI.1.

Statements/Functions List

$+, -, *, /, >$
GOSUB, IF/THEN

Variables List

A(I), C, N, X, X1, X2, X3, Y, Z, Z1

Variables Passed to Subroutine

X

Table VI.18 Statements, functions, and variables used in the logarithm subroutines shown in programs VI.5 and VI.6 (LOG(X), LN(X)). The items shown in table VI.15 should be included in the list below.

```
43259 REM LOG BASE TEN SERIES SUBROUTINE (LOG(X))                    (a)
43260 X1=1
43261 C=-1
43262 X2=10
43263 X3=X
43264 IF X>=1 THEN GOTO 43268
43265 X=1/X
43266 X1=-1
43267 REM REDUCE RANGE
43268 X=X2*X
43269 C=C+1
43270 X=X/X2
43271 IF X>X2 THEN GOTO 43269
43272 Z1=(X-3.162277660)/(X+3.162277660)
43273 Z=Z1*Z1
43274 GOSUB 43340
43275 GOSUB 43200
43276 Y=X1*(C+Z1*Y+.5)
43277 X=X3
43278 RETURN


43339 REM LOG BASE TEN SERIES COEFFICIENTS (LOGDATA)                 (b)
43340 N=9
43341 A(0)=.8685889644
43342 A(1)=.2895297117
43343 A(2)=.1737120608
43344 A(3)=.1242584742
43345 A(4)=.0939080460
43346 A(5)=.1009301264
43347 A(6)=-.0439630355
43348 A(7)=.3920576195
43349 A(8)=-.5170494708
43350 A(9)=.4915571108
43351 RETURN
```

Program VI.5: *Logarithm to the base ten series-approximation subroutine 5a (LOG(X)) with its associated coefficients (5b). Also uses program VI.1.*

```
43279 REM NATURAL LOGARITHM SERIES SUBROUTINE (LN(X))         Program VI.6:
43280 X1=1
43281 C=-1                                                           (a)
43282 X2=2.718281828459045
43283 X3=X
43284 IF X>=1 THEN GOTO 43288
43285 X=1/X
43286 X1=-1
43287 REM REDUCE RANGE
43288 X=X2*X
43289 C=C+1
```

```
43290 X=X/X2
43291 IF X>X2 THEN GOTO 43289
43292 Z1=(X-1.6487212707)/(X+1.6487212707)
43293 Z=Z1*Z1
43294 GOSUB 43360
43295 GOSUB 43200
43296 Y=X1*(C+Z1*Y+.5)
43297 X=X3
43298 RETURN
```

(b)
```
43359 REM NATURAL LOGARITHM SERIES COEFFICIENTS (LNDATA)
43360 N=9
43361 A(0)=2
43362 A(1)=.66666672443
43363 A(2)=.3999895288
43364 A(3)=.286436047
43365 A(4)=.197959107
43366 A(5)=.6283533
43367 A(6)=-4.54692
43368 A(7)=28.117
43369 A(8)=-86.42
43370 A(9)=106.1
43371 RETURN
```

Program VI.6: *Natural logarithm series-approximation subroutine 6a (LN(X)) with its associated coefficients (6b). Also uses program VI.1.*

Statements/Functions List

$-, *, /, <$

ABS, FOR/NEXT, GOSUB, IF/THEN, INT

Variables List

A(I), I, N, X, X1, X2, X3, Y, Z

Variables Passed to Subroutine

X

Table VI.19 *Statements, functions, and variables used in the exponentiation subroutines shown in programs VI.7 and VI.8 (TENPOW, EXP(X)). The items in table VI.15 should be included in the list below.*

```
43449 REM POWER OF TEN SERIES SUBROUTINE (TENPOW)           (a)
43450 X1=1
43451 X3=X
43452 IF X<0 THEN X1=-1
43453 X=ABS(X)
43454 REM REDUCE RANGE
43455 X2=INT(X)
43456 X=X-X2
43457 REM GET COEFFICIENTS
43458 GOSUB 43400
43459 Z=X
43460 REM SUM SERIES
43461 GOSUB 43200
43462 Y=Y*Y
43463 FOR I=1 TO X2
43464 Y=Y*10
43465 NEXT I
43466 IF X1<0 THEN Y=1/Y
43467 X=X3
43468 RETURN


43399 REM POWER OF TEN SERIES COEFFICIENTS (TENDATA)        (b)
43400 N=9
43401 A(0)=1
43402 A(1)=1.1512925485
43403 A(2)=.6627373050
43404 A(3)=.2543345675
43405 A(4)=.0732032741
43406 A(5)=.0168603036
43407 A(6)=.0032196227
43408 A(7)=.0005547662
43409 A(8)=.0000573305
43410 A(9)=.0000179419
43411 RETURN
```

Program VI.7: *Power of ten series-approximation subroutine 7a (TENPOW) with its associated coefficients (7b). Also uses program VI.1.*

```
43469 REM EXPONENT SERIES SUBROUTINE (EXP(X))        Program VI.8:
43470 X1=1
43471 X3=X                                               (a)
43472 IF X<0 THEN X1=-1
43473 X=ABS(X)
43474 REM REDUCE RANGE
43475 X2=INT(X)
43476 X=X-X2
43477 REM GET COEFFICIENTS
43478 GOSUB 43380
```

```
43479 Z=X
43480 REM SUM SERIES
43481 GOSUB 43200
43482 FOR I=1 TO X2
43483 Y=Y*2.718281828459045
43484 NEXT I
43485 IF X1<0 THEN Y=1/Y
43486 X=X3
43487 RETURN
```

(b)
```
43379 REM POWER OF E SERIES COEFFICIENTS (EXPDATA)
43380 N=8
43381 A(0)=1
43382 A(1)=.99999999668
43383 A(2)=.49999995173
43384 A(3)=.16666704243
43385 A(4)=.04166685027
43386 A(5)=.00832672635
43387 A(6)=.00140836136
43388 A(7)=.00017358267
43389 A(8)=.00003931683
43390 RETURN
```

Program VI.8: *Power of e series-approximation subroutine 8a (EXP(X)) with its associated coefficients (8b). Also uses program VI.1.*

```
REM SINE AND COSINE APPROXIMATION DEMONSTRATION
PRINT"   X          SIN(X)              DELTA         COS(X)           DELTA"
PRINT" ---         --------            -------       -------         -------"
PRINT
FOR X=-5 TO 5 STEP .2
GOSUB 43210
PRINT %4F1,X,%19F14,Y,%13E2,(SIN(X)-Y),
GOSUB 43225
PRINT %26F14,Y,%13E2,(COS(X)-Y)
NEXT X
END
```

Program VI.9: *Program to demonstrate the use of the sine and cosine subroutines (programs VI.2 and VI.3). See listings VI.1a and VI.1b for results.*

X	SIN(X)	DELTA	COS(X)	DELTA
-5.0	.95892432000000	-2.00E-08	.28366210000000	1.20E-07
-4.8	.99616457000000	3.00E-08	.08749890000000	1.10E-07
-4.6	.99369092000000	8.00E-08	-.11215250000000	.00E+00
-4.4	.95160206000000	4.00E-08	-.30733290000000	6.00E-08
-4.2	.87157576000000	-6.00E-08	-.49026090000000	1.00E-07
-4.0	.75680243000000	7.00E-08	-.65364370000000	1.00E-07
-3.8	.61185784000000	6.00E-08	-.79096770000000	1.00E-08
-3.6	.44252042000000	3.00E-08	-.89675840000000	.00E+00
-3.4	.25554105000000	6.00E-08	-.96679820000000	.00E+00
-3.2	.05837409500000	4.80E-08	-.99829480000000	1.00E-07
-3.0	-.14112005000000	5.00E-08	-.98999250000000	.00E+00
-2.8	-.33498821000000	6.00E-08	-.94222230000000	-1.00E-07
-2.6	-.51550142000000	5.00E-08	-.85688870000000	-1.00E-07
-2.4	-.67546325000000	7.00E-08	-.73739370000000	-3.00E-08
-2.2	-.80849641000000	1.00E-08	-.58850110000000	-4.00E-08
-2.0	-.90929742000000	-8.00E-08	-.41614680000000	-6.00E-08
-1.8	-.97384765000000	5.00E-08	-.22720200000000	-1.20E-07
-1.6	-.99957364000000	4.00E-08	-.02919950000000	-4.90E-08
-1.4	-.98544978000000	8.00E-08	.16996710000000	2.00E-08
-1.2	-.93203904000000	-6.00E-08	.36235780000000	-7.00E-08
-1.0	-.84147100000000	.00E+00	.54030230000000	-1.00E-08
-.8	-.71735608000000	-2.00E-08	.69670670000000	-1.00E-08
-.6	-.56464248000000	.00E+00	.82533560000000	.00E+00
-.4	-.38941836000000	1.00E-08	.92106100000000	.00E+00
-.2	-.19866934000000	1.00E-08	.98006660000000	.00E+00
.0	.00000000000000	.00E+00	1.00000000000000	.00E+00
.2	.19866934000000	-1.00E-08	.98006660000000	.00E+00
.4	.38941836000000	-1.00E-08	.92106100000000	.00E+00
.6	.56464248000000	.00E+00	.82533560000000	2.00E-08
.8	.71735608000000	2.00E-08	.69670670000000	3.00E-08
1.0	.84147100000000	.00E+00	.54030230000000	3.00E-08
1.2	.93203904000000	6.00E-08	.36235780000000	-2.00E-08
1.4	.98544978000000	-8.00E-08	.16996710000000	7.00E-08
1.6	.99957364000000	-4.00E-08	-.02919950000000	5.00E-09
1.8	.97384765000000	-5.00E-08	-.22720200000000	-7.00E-08
2.0	.90929742000000	8.00E-08	-.41614680000000	-2.00E-08
2.2	.80849641000000	-1.00E-08	-.58850110000000	.00E+00
2.4	.67546325000000	-7.00E-08	-.73739370000000	.00E+00
2.6	.51550142000000	-5.00E-08	-.85688870000000	.00E+00
2.8	.33498821000000	-6.00E-08	-.94222230000000	.00E+00
3.0	.14112005000000	-5.00E-08	-.98999250000000	.00E+00
3.2	-.05837409500000	-4.80E-08	-.99829480000000	1.00E-07
3.4	-.25554105000000	-6.00E-08	-.96679820000000	.00E+00
3.6	-.44252042000000	-3.00E-08	-.89675840000000	.00E+00
3.8	-.61185784000000	-6.00E-08	-.79096770000000	-3.00E-08
4.0	-.75680243000000	-7.00E-08	-.65364370000000	5.00E-08
4.2	-.87157576000000	6.00E-08	-.49026090000000	5.00E-08
4.4	-.95160206000000	-4.00E-08	-.30733290000000	.00E+00
4.6	-.99369092000000	-8.00E-08	-.11215250000000	-6.00E-08
4.8	-.99616457000000	-3.00E-08	.08749890000000	5.70E-08
5.0	-.95892432000000	2.00E-08	.28366210000000	6.00E-08

Listing VI.1a: *Comparison of the sine and cosine approximations with the function values calculated using eight-digit North Star BASIC.*

X	SIN(X)	DELTA	COS(X)	DELTA
-5.0	.95892427467039	-7.29E-12	.28366218545400	9.23E-12
-4.8	.99616460899615	-1.60E-10	.08749898336840	7.10E-11
-4.6	.99369100374918	-1.16E-10	-.11215252687360	-6.15E-11
-4.4	.95160207389397	-4.47E-12	-.30733286997220	-6.22E-12
-4.2	.87157577241354	6.00E-14	-.49026082134080	1.00E-13
-4.0	.75680249530793	.00E+00	-.65364362086360	-1.00E-14
-3.8	.61185789094271	1.00E-14	-.79096771191450	8.00E-14
-3.6	.44252044329487	-1.00E-14	-.89675841633420	1.00E-13
-3.4	.25554110202681	2.00E-14	-.96679819257950	.00E+00
-3.2	.05837414342757	6.00E-15	-.99829477579480	1.00E-13
-3.0	-.14112000805987	1.00E-14	-.98999249660050	1.00E-13
-2.8	-.33498815015591	1.00E-14	-.94222234066870	1.00E-13
-2.6	-.51550137182148	2.00E-14	-.85688875336890	.00E+00
-2.4	-.67546318055116	1.00E-14	-.73739371554130	6.00E-14
-2.2	-.80849640381961	2.00E-14	-.58850111725530	-4.00E-14
-2.0	-.90929742682584	1.40E-13	-.41614683654680	-3.40E-13
-1.8	-.97384763089867	2.05E-11	-.22720209467260	-2.05E-11
-1.6	-.99957360337284	3.31E-10	-.02919952221900	-8.23E-11
-1.4	-.98544973003946	5.10E-11	.16996714286190	3.83E-11
-1.2	-.93203908596840	1.20E-12	.36235775447470	1.98E-12
-1.0	-.84147098480790	.00E+00	.54030230586820	-6.00E-14
-.8	-.71735609089952	.00E+00	.69670670934720	-3.00E-14
-.6	-.56464247339502	-2.00E-14	.82533561490970	-2.00E-14
-.4	-.38941834230864	-1.00E-14	.92106099400290	.00E+00
-.2	-.19866933079506	.00E+00	.98006657784130	-1.00E-13
.0	.00000000000000	.00E+00	1.00000000000000	.00E+00
.2	.19866933079506	.00E+00	.98006657784130	-1.00E-13
.4	.38941834230864	1.00E-14	.92106099400290	.00E+00
.6	.56464247339502	2.00E-14	.82533561490970	-2.00E-14
.8	.71735609089952	.00E+00	.69670670934720	-4.00E-14
1.0	.84147098480790	.00E+00	.54030230586820	-7.00E-14
1.2	.93203908596840	-1.20E-12	.36235775447470	1.97E-12
1.4	.98544973003946	-5.10E-11	.16996714286190	3.83E-11
1.6	.99957360337284	-3.31E-10	-.02919952221900	-8.23E-11
1.8	.97384763089867	-2.05E-11	-.22720209467260	-2.05E-11
2.0	.90929742682584	-1.40E-13	-.41614683654680	-3.50E-13
2.2	.80849640381961	-2.00E-14	-.58850111725530	-5.00E-14
2.4	.67546318055116	-1.00E-14	-.73739371554130	5.00E-14
2.6	.51550137182148	-2.00E-14	-.85688875336890	.00E+00
2.8	.33498815015591	-1.00E-14	-.94222234066870	1.00E-13
3.0	.14112000805987	-1.00E-14	-.98999249660050	1.00E-13
3.2	-.05837414342757	-6.00E-15	-.99829477579480	1.00E-13
3.4	-.25554110202681	-2.00E-14	-.96679819257950	.00E+00
3.6	-.44252044329487	1.00E-14	-.89675841633420	1.00E-13
3.8	-.61185789094271	-1.00E-14	-.79096771191450	8.00E-14
4.0	-.75680249530793	.00E+00	-.65364362086360	-1.00E-14
4.2	-.87157577241354	-6.00E-14	-.49026082134080	1.00E-13
4.4	-.95160207389397	4.47E-12	-.30733286997220	-6.22E-12
4.6	-.99369100374918	1.16E-10	-.11215252687360	-6.15E-11
4.8	-.99616460899615	1.60E-10	.08749898336840	7.10E-11
5.0	-.95892427467039	7.29E-12	.28366218545400	9.23E-12

Listing VI.1b: *Comparison of the sine and cosine approximations with the functions calculated using fourteen-digit North Star BASIC.*

```
REM ARCTANGENT APPROXIMATION DEMONSTRATION
PRINT"   X                  ARCTANGENT                    DELTA"
PRINT"---              ------------               -------"
PRINT
FOR X=-5 TO 5 STEP .2
GOSUB 43245
PRINT %4F1,X,%23F14,Y,%14E2,(ATN(X)-Y)
NEXT X
END
```

Program VI.10: *Program to demonstrate the use of the arctangent subroutine (program VI.4). See results in listings VI.2a and VI.2b.*

X	ARCTANGENT	DELTA
---	------------	-------
-5.0	-1.37340080000000	1.00E-07
-4.8	-1.36540090000000	.00E+00
-4.6	-1.35673560000000	.00E+00
-4.4	-1.34731980000000	1.00E-07
-4.2	-1.33705310000000	.00E+00
-4.0	-1.32581770000000	1.00E-07
-3.8	-1.31347260000000	.00E+00
-3.6	-1.29984950000000	1.00E-07
-3.4	-1.28474490000000	.00E+00
-3.2	-1.26791140000000	.00E+00
-3.0	-1.24904580000000	1.00E-07
-2.8	-1.22777240000000	.00E+00
-2.6	-1.20362250000000	.00E+00
-2.4	-1.17600520000000	.00E+00
-2.2	-1.14416880000000	.00E+00
-2.0	-1.10714870000000	.00E+00
-1.8	-1.06369780000000	.00E+00
-1.6	-1.01219700000000	.00E+00
-1.4	-.95054685000000	5.00E-08
-1.2	-.87605805000000	5.00E-08
-1.0	-.78539816000000	6.00E-08
-.8	-.67474094000000	.00E+00
-.6	-.54041948000000	-2.00E-08
-.4	-.38050636000000	-2.00E-08
-.2	-.19739556000000	.00E+00
.0	.00000000000000	.00E+00
.2	.19739556000000	.00E+00
.4	.38050636000000	2.00E-08
.6	.54041948000000	2.00E-08
.8	.67474094000000	.00E+00
1.0	.78539816000000	-6.00E-08
1.2	.87605805000000	-5.00E-08
1.4	.95054685000000	-5.00E-08
1.6	1.01219700000000	.00E+00
1.8	1.06369780000000	.00E+00
2.0	1.10714870000000	.00E+00
2.2	1.14416880000000	.00E+00
2.4	1.17600520000000	.00E+00
2.6	1.20362250000000	.00E+00
2.8	1.22777240000000	.00E+00
3.0	1.24904580000000	-1.00E-07
3.2	1.26791140000000	.00E+00
3.4	1.28474490000000	.00E+00
3.6	1.29984950000000	-1.00E-07
3.8	1.31347260000000	.00E+00
4.0	1.32581770000000	-1.00E-07
4.2	1.33705310000000	.00E+00
4.4	1.34731980000000	-1.00E-07
4.6	1.35673560000000	.00E+00
4.8	1.36540090000000	.00E+00
5.0	1.37340080000000	-1.00E-07

Listing VI.2a: *Comparison of the arctangent approximation with the values calculated using eight-digit North Star Basic.*

X	ARCTANGENT	DELTA
---	-------------	-------
-5.0	-1.37340076685720	-7.38E-11
-4.8	-1.36540093754230	-6.27E-11
-4.6	-1.35673564320590	-4.15E-11
-4.4	-1.34731972567330	-1.59E-11
-4.2	-1.33705314598750	6.70E-12
-4.0	-1.32581766376100	1.85E-11
-3.8	-1.31347261192940	1.45E-11
-3.6	-1.29984947655150	-5.70E-12
-3.4	-1.28474488514140	-3.36E-11
-3.2	-1.26791145844200	-5.28E-11
-3.0	-1.24904577238500	-4.13E-11
-2.8	-1.22777238634850	1.55E-11
-2.6	-1.20362249296850	1.06E-10
-2.4	-1.17600520712200	1.71E-10
-2.2	-1.14416883371800	1.18E-10
-2.0	-1.10714871783180	-7.43E-11
-1.8	-1.06369782240600	-1.54E-10
-1.6	-1.01219701144450	1.58E-10
-1.4	-.95054684082519	-8.06E-11
-1.2	-.87605805060835	2.66E-10
-1.0	-.78539816339745	2.92E-10
-.8	-.67474094220936	-5.39E-11
-.6	-.54041950027882	-3.48E-11
-.4	-.38050637714984	-1.01E-10
-.2	-.19739555985230	-1.16E-11
.0	.00000000000000	.00E+00
.2	.19739555985230	1.16E-11
.4	.38050637714984	1.01E-10
.6	.54041950027882	3.48E-11
.8	.67474094220936	5.39E-11
1.0	.78539816339745	-2.92E-10
1.2	.87605805060835	-2.66E-10
1.4	.95054684082519	8.06E-11
1.6	1.01219701144450	-1.58E-10
1.8	1.06369782240600	1.54E-10
2.0	1.10714871783180	7.43E-11
2.2	1.14416883371800	-1.18E-10
2.4	1.17600520712200	-1.71E-10
2.6	1.20362249296850	-1.06E-10
2.8	1.22777238634850	-1.55E-11
3.0	1.24904577238500	4.13E-11
3.2	1.26791145844200	5.28E-11
3.4	1.28474488514140	3.36E-11
3.6	1.29984947655150	5.70E-12
3.8	1.31347261192940	-1.45E-11
4.0	1.32581766376100	-1.85E-11
4.2	1.33705314598750	-6.70E-12
4.4	1.34731972567330	1.59E-11
4.6	1.35673564320590	4.15E-11
4.8	1.36540093754230	6.27E-11
5.0	1.37340076685720	7.38E-11

Listing VI.2b: *Comparison of the arctangent approximation with the functions calculating using fourteen-digit North Star BASIC.*

```
REM BASE 10 LOGARITHM AND EXPONENT APPROXIMATION DEMONSTRATION
PRINT "    X           LOG(X)           DELTA         10^(LOG(X))        DELTA"
PRINT "   ---      --------------      -------      --------------     --------"
FOR X=.1 TO 5 STEP .1
GOSUB 43260
PRINT %3F1," ",X,%19F14,Y,%13E2,(.434294481903251*LOG(X)-Y),
F=X
X=Y
GOSUB 43450
```

```
PRINT %25F14,Y,%14E2,(F-Y)
X=F
NEXT X
END
```

Program VI.11: *Program to demonstrate the use of the base 10 log and power subroutines (programs VI.5 and VI.7). See listings VI.3a and VI.3b for results.*

X	LOG(X)	DELTA	10^(LOG(X))	DELTA
.1	-.99999995000000	.00E+00	.100000020000000	-2.00E-08
.2	-.69897000000000	.00E+00	.200000010000000	-1.00E-08
.3	-.52287874000000	4.00E-08	.300000020000000	-2.00E-08
.4	-.39794000000000	-1.00E-08	.400000020000000	-2.00E-08
.5	-.30102999000000	.00E+00	.500000050000000	-5.00E-08
.6	-.22184875000000	.00E+00	.600000060000000	-6.00E-08
.7	-.15490195000000	-1.00E-08	.700000010000000	-1.00E-08
.8	-.09691001000000	-2.00E-09	.800000000000000	.00E+00
.9	-.04575749000000	-2.00E-09	.899999930000000	7.00E-08
1.0	.00000000000000	.00E+00	1.000000000000000	.00E+00
1.1	.04139268000000	5.00E-09	1.099999990000000	1.00E-07
1.2	.07918124000000	7.00E-09	1.200000000000000	.00E+00
1.3	.11394335000000	.00E+00	1.299999990000000	1.00E-07
1.4	.14612803000000	1.00E-08	1.399999990000000	1.00E-07
1.5	.17609125000000	1.00E-08	1.500000010000000	-1.00E-07
1.6	.20411998000000	.00E+00	1.600000010000000	-1.00E-07
1.7	.23044892000000	.00E+00	1.700000000000000	.00E+00
1.8	.25527250000000	1.00E-08	1.800000000000000	.00E+00
1.9	.27875360000000	.00E+00	1.900000010000000	-1.00E-07
2.0	.30102999000000	.00E+00	1.999999980000000	2.00E-07
2.1	.32221929000000	1.00E-08	2.100000010000000	-1.00E-07
2.2	.34242268000000	.00E+00	2.200000000000000	.00E+00
2.3	.36172783000000	.00E+00	2.300000000000000	.00E+00
2.4	.38021124000000	1.00E-08	2.399999990000000	1.00E-07
2.5	.39794000000000	1.00E-08	2.499999990000000	1.00E-07
2.6	.41497334000000	.00E+00	2.599999980000000	2.00E-07
2.7	.43136376000000	1.00E-08	2.699999980000000	2.00E-07
2.8	.44715803000000	-1.00E-08	2.800000020000000	-2.00E-07
2.9	.46239799000000	-1.00E-08	2.899999990000000	1.00E-07
3.0	.47712125000000	1.00E-08	3.000000000000000	.00E+00
3.1	.49136169000000	.00E+00	3.100000000000000	.00E+00
3.2	.50514997000000	5.00E-08	3.200000010000000	-1.00E-07
3.3	.51851393000000	-2.00E-08	3.300000000000000	.00E+00
3.4	.53147891000000	-1.00E-08	3.399999970000000	3.00E-07
3.5	.54406804000000	-3.00E-08	3.500000000000000	.00E+00
3.6	.55630250000000	-2.00E-08	3.600000000000000	.00E+00
3.7	.56820172000000	-1.00E-08	3.700000000000000	.00E+00
3.8	.57978359000000	2.00E-08	3.799999970000000	3.00E-07
3.9	.59106460000000	-2.00E-08	3.899999970000000	3.00E-07
4.0	.60205998000000	3.00E-08	4.000000000000000	.00E+00
4.1	.61278385000000	-7.00E-08	4.099999970000000	3.00E-07
4.2	.62324929000000	3.00E-08	4.199999980000000	2.00E-07
4.3	.63346845000000	-5.00E-08	4.299999990000000	1.00E-07
4.4	.64345267000000	3.00E-08	4.400000000000000	.00E+00
4.5	.65321251000000	.00E+00	4.499999980000000	2.00E-07
4.6	.66275783000000	-5.00E-08	4.600000020000000	-2.00E-07
4.7	.67209785000000	4.00E-08	4.699999980000000	2.00E-07
4.8	.68124123000000	-5.00E-08	4.799999990000000	1.00E-07
4.9	.69019607000000	5.00E-08	4.899999970000000	3.00E-07
5.0	.69897000000000	.00E+00	4.999999970000000	3.00E-07

Listing VI.3a: *Comparison of the base 10 log and power approximations with the functions calculated using eight-digit North Star BASIC.*

Listing VI.3b:
Comparison of the base 10 log and power approximations with the functions calculated using fourteen-digit North Star BASIC.

X	LOG(X)	DELTA	10^(LOG(X))	DELTA
.1	-.99999999999224	-7.80E-12	.10000000001243	-1.24E-11
.2	-.69897000436469	2.87E-11	.19999999987898	1.21E-10
.3	-.52287874532006	3.97E-11	.29999999988763	1.12E-10
.4	-.39794000861715	-5.49E-11	.39999999996726	3.27E-11
.5	-.30102999568156	1.76E-11	.49999999976595	2.34E-10
.6	-.22184874956924	-4.71E-11	.59999999963081	3.69E-10
.7	-.15490196003926	5.35E-11	.69999999933420	6.66E-10
.8	-.09691001300632	-1.74E-12	.79999999948666	5.13E-10
.9	-.04575749056130	6.28E-13	.89999999973676	2.63E-10
1.0	.00000000005399	-5.40E-11	1.00000000012440	-1.24E-10
1.1	.04139268516766	-9.44E-12	1.10000000030760	-3.08E-10
1.2	.07918124602331	2.43E-11	1.20000000057640	-5.76E-10
1.3	.11394335233617	-2.93E-11	1.30000000103540	-1.04E-09
1.4	.14612803573661	-5.84E-11	1.40000000133440	-1.33E-09
1.5	.17609125907835	-2.27E-11	1.50000000131590	-1.32E-09
1.6	.20411998262675	2.92E-11	1.60000000113300	-1.13E-09
1.7	.23044892132933	4.89E-11	1.70000000098950	-9.90E-10
1.8	.25527250507182	3.15E-11	1.80000000095370	-9.54E-10
1.9	.27875360095186	9.70E-13	1.90000000096620	-9.66E-10
2.0	.30102999568156	-1.76E-11	2.00000000093620	-9.36E-10
2.1	.32221929474821	-1.43E-11	2.10000000081890	-8.19E-10
2.2	.34242268081652	5.68E-12	2.20000000063180	-6.32E-10
2.3	.36172783598738	3.02E-11	2.30000000043190	-4.32E-10
2.4	.38021124166320	4.84E-11	2.40000000027770	-2.78E-10
2.5	.39794000861715	5.49E-11	2.50000000020460	-2.05E-10
2.6	.41497334792093	4.99E-11	2.60000000021800	-2.18E-10
2.7	.43136376412178	3.72E-11	2.70000000030070	-3.01E-10
2.8	.44715803132072	2.15E-11	2.80000000042590	-4.26E-10
2.9	.46239799789260	6.34E-12	2.90000000056930	-5.69E-10
3.0	.47712125472619	-6.53E-12	3.00000000071710	-7.17E-10
3.1	.49136169385144	-1.72E-11	3.10000000086630	-8.66E-10
3.2	.50514997834656	-2.67E-11	3.20000000102100	-1.02E-09
3.3	.51851393991419	-3.63E-11	3.30000000118880	-1.19E-09
3.4	.53147891708925	-4.71E-11	3.40000000137530	-1.38E-09
3.5	.54406804440915	-5.89E-11	3.50000000158130	-1.58E-09
3.6	.55630250083858	-7.13E-11	3.60000000180160	-1.80E-09
3.7	.56820172415005	-8.31E-11	3.70000000202460	-2.02E-09
3.8	.57978359670955	-9.28E-11	3.80000000223710	-2.24E-09
3.9	.59106460712555	-9.91E-11	3.90000000242530	-2.43E-09
4.0	.60205999142909	-1.01E-10	4.00000000257840	-2.58E-09
4.1	.61278385681841	-9.87E-11	4.10000000269000	-2.69E-09
4.2	.62324929048992	-9.20E-11	4.20000000275960	-2.76E-09
4.3	.63346845566162	-8.20E-11	4.30000000279250	-2.79E-09
4.4	.64345267655620	-7.00E-11	4.40000000279860	-2.80E-09
4.5	.65321251383278	-5.75E-11	4.50000000279230	-2.79E-09
4.6	.66275783172734	-4.58E-11	4.60000000278770	-2.79E-09
4.7	.67209785797201	-3.63E-11	4.70000000279850	-2.80E-09
4.8	.68124123740557	-3.00E-11	4.80000000283740	-2.84E-09
4.9	.69019608005592	-2.74E-11	4.90000000291150	-2.91E-09
5.0	.69897000436469	-2.87E-11	5.00000000302540	-3.03E-09

```
REM NATURAL LOGARITHM AND EXPONENT APPROXIMATION DEMONSTRATION
PRINT "    X       LN(X)             DELTA          EXP(LN(X))        DELTA"
PRINT "   ---   -------           -------         ------------       -------"
FOR X=.1 TO 5 STEP .1
GOSUB 43280
PRINT %3F1," ",X,%19F14,Y,%13E2,(LOG(X)-Y),
F=X
X=Y
GOSUB 43470
PRINT %25F14,Y,%14E2,(F-Y)
X=F
NEXT X
END
```

Program VI.12: *Program to demonstrate the use of the logarithm and exponent approximation subroutines (programs VI.6 and VI.8). See results shown in listings VI.4a and VI.4b.*

X	LN(X)	DELTA	EXP(LN(X))	DELTA
.1	-2.30258510000000	.00E+00	.10000000000000	.00E+00
.2	-1.60943790000000	.00E+00	.20000000000000	.00E+00
.3	-1.20397280000000	1.00E-07	.30000000000000	.00E+00
.4	-.91629071000000	-3.00E-08	.40000000000000	.00E+00
.5	-.69314716000000	-1.00E-08	.50000003000000	-3.00E-08
.6	-.51082563000000	.00E+00	.59999999000000	1.00E-08
.7	-.35667491000000	-4.00E-08	.70000001000000	-1.00E-08
.8	-.22314355000000	.00E+00	.80000000000000	.00E+00
.9	-.10536049000000	-3.00E-08	.90000001000000	-1.00E-08
1.0	-.00000002000000	2.00E-08	1.00000000000000	.00E+00
1.1	.09531017000000	1.00E-08	1.10000000000000	.00E+00
1.2	.18232155000000	1.00E-08	1.20000000000000	.00E+00
1.3	.26236426000000	.00E+00	1.30000000000000	.00E+00
1.4	.33647222000000	3.00E-08	1.40000000000000	.00E+00
1.5	.40546509000000	2.00E-08	1.50000000000000	.00E+00
1.6	.47000361000000	2.00E-08	1.60000000000000	.00E+00
1.7	.53062823000000	2.00E-08	1.70000000000000	.00E+00
1.8	.58778665000000	3.00E-08	1.80000000000000	.00E+00
1.9	.64185387000000	2.00E-08	1.90000000000000	.00E+00
2.0	.69314716000000	1.00E-08	1.99999990000000	1.00E-07
2.1	.74193734000000	2.00E-08	2.10000000000000	.00E+00
2.2	.78845734000000	2.00E-08	2.20000000000000	.00E+00
2.3	.83290912000000	-1.00E-08	2.30000000000000	.00E+00
2.4	.87546872000000	4.00E-08	2.39999990000000	1.00E-07
2.5	.91629071000000	3.00E-08	2.50000000000000	.00E+00
2.6	.95551144000000	-1.00E-08	2.60000000000000	.00E+00
2.7	.99325177000000	2.00E-08	2.70000000000000	.00E+00
2.8	1.02961930000000	1.00E-07	2.79999960000000	4.00E-07
2.9	1.06471070000000	.00E+00	2.89999980000000	2.00E-07
3.0	1.09861220000000	1.00E-07	2.99999960000000	4.00E-07
3.1	1.13140210000000	.00E+00	3.10000010000000	-1.00E-07
3.2	1.16315080000000	1.00E-07	3.19999990000000	1.00E-07
3.3	1.19392250000000	-1.00E-07	3.30000010000000	-1.00E-07
3.4	1.22377540000000	.00E+00	3.40000000000000	.00E+00
3.5	1.25276300000000	-1.00E-07	3.50000010000000	-1.00E-07
3.6	1.28093380000000	.00E+00	3.59999970000000	3.00E-07
3.7	1.30833280000000	.00E+00	3.69999990000000	1.00E-07
3.8	1.33500110000000	.00E+00	3.80000000000000	.00E+00
3.9	1.36097650000000	.00E+00	3.89999960000000	4.00E-07
4.0	1.38629430000000	1.00E-07	3.99999980000000	2.00E-07
4.1	1.41098690000000	-1.00E-07	4.09999970000000	3.00E-07
4.2	1.43508450000000	1.00E-07	4.19999980000000	2.00E-07
4.3	1.45861510000000	-2.00E-07	4.30000020000000	-2.00E-07
4.4	1.48160460000000	.00E+00	4.40000040000000	-4.00E-07
4.5	1.50407740000000	.00E+00	4.50000000000000	.00E+00
4.6	1.52605630000000	-1.00E-07	4.59999990000000	1.00E-07
4.7	1.54756240000000	2.00E-07	4.69999950000000	5.00E-07
4.8	1.56861590000000	-1.00E-07	4.79999990000000	1.00E-07
4.9	1.58923520000000	1.00E-07	4.89999980000000	2.00E-07
5.0	1.60943790000000	.00E+00	4.99999990000000	1.00E-07

Listing VI.4a: *Comparison of the natural log and exponent approximations with the functions calculated using eight-digit North Star BASIC.*

X	LN(X)	DELTA	EXP(LN(X))	DELTA
.1	-2.30258509298890	-5.20E-12	.10000000000754	-7.54E-12
.2	-1.60943791243950	5.40E-12	.20000000004812	-4.81E-11
.3	-1.20397280432100	-4.90E-12	.30000000010721	-1.07E-10
.4	-.91629073190310	2.89E-11	.40000000008832	-8.83E-11
.5	-.69314718056572	5.77E-12	.50000000003408	-3.41E-11
.6	-.51082562376610	1.00E-13	.60000000016850	-1.69E-10
.7	-.35667494393142	-7.31E-12	.70000000002917	-2.92E-11
.8	-.22314355131241	-1.80E-12	.80000000023443	-2.34E-10
.9	-.10536051562715	-3.07E-11	.90000000042129	-4.21E-10
1.0	.00000000000101	-1.01E-12	1.00000000000100	-1.00E-12
1.1	.09531017977391	3.04E-11	1.09999999951310	4.87E-10
1.2	.18232155678329	1.07E-11	1.19999999949260	5.07E-10
1.3	.26236426446610	1.40E-12	1.29999999978280	2.17E-10
1.4	.33647223661385	7.37E-12	1.39999999994140	5.86E-11
1.5	.40546510810427	3.89E-12	1.49999999986010	1.40E-10
1.6	.47000362924563	1.00E-13	1.59999999965140	3.49E-10
1.7	.53062825106245	-2.90E-13	1.69999999949530	5.05E-10
1.8	.58778666490558	-3.45E-12	1.79999999950690	4.93E-10
1.9	.64185388617979	-7.39E-12	1.89999999966730	3.33E-10
2.0	.69314718056572	-5.77E-12	1.99999999986370	1.36E-10
2.1	.74193734473066	-1.28E-12	2.09999999997990	2.01E-11
2.2	.78845736036785	-3.59E-12	2.19999999995870	4.13E-11
2.3	.83290912295095	-1.59E-11	2.29999999981140	1.89E-10
2.4	.87546873738251	-2.86E-11	2.39999999960450	3.96E-10
2.5	.91629073190310	-2.89E-11	2.49999999944800	5.52E-10
2.6	.95551144504231	-1.49E-11	2.59999999947390	5.26E-10
2.7	.99325177301059	-3.00E-13	2.69999999977500	2.25E-10
2.8	1.02961941717320	8.00E-12	2.79999999962170	3.78E-10
2.9	1.06471073696910	2.33E-11	2.89999999906760	9.32E-10
3.0	1.09861228863750	3.06E-11	2.99999999864430	1.36E-09
3.1	1.13140211146420	2.69E-11	3.09999999845790	1.54E-09
3.2	1.16315080978860	1.71E-11	3.19999999850350	1.50E-09
3.3	1.19392246846510	7.40E-12	3.29999999871570	1.28E-09
3.4	1.22377543162040	1.60E-12	3.39999999901100	9.89E-10
3.5	1.25276296849450	9.00E-13	3.49999999931510	6.85E-10
3.6	1.28093384545910	3.00E-12	3.59999999957350	4.27E-10
3.7	1.30833281964440	5.80E-12	3.69999999975270	2.47E-10
3.8	1.33500106672500	7.30E-12	3.79999999983880	1.61E-10
3.9	1.36097655312840	7.20E-12	3.89999999983020	1.70E-10
4.0	1.38629436111440	5.60E-12	3.99999999973820	2.62E-10
4.1	1.41098697370680	3.40E-12	4.09999999957910	4.21E-10
4.2	1.43508452528780	1.50E-12	4.19999999937730	6.23E-10
4.3	1.45861502269920	4.00E-13	4.29999999915860	8.41E-10
4.4	1.48160454092430	.00E+00	4.39999999894860	1.05E-09
4.5	1.50407739677640	-2.00E-13	4.49999999877180	1.23E-09
4.6	1.52605630349520	-1.00E-13	4.59999999864640	1.35E-09
4.7	1.54756250871680	-8.00E-13	4.69999999858540	1.41E-09
4.8	1.56861591791580	-1.90E-12	4.79999999859310	1.41E-09
4.9	1.58923520512020	-3.60E-12	4.89999999866670	1.33E-09
5.0	1.60943791243950	-5.40E-12	4.99999999879700	1.20E-09

Listing VI.4b: *Comparison of the natural log and exponent approximations with the functions calculated using fourteen-digit North Star BASIC.*

VI.5 Extensions

The subroutines presented in the previous section can be expanded in utility through the use of table VI.20. Reference 30 contains further relationships which permit the approximation of a wide class of functions using the basic sine, logarithm, and exponent routines. However, in many cases, faster execution can be attained by series approximation of the desired function rather than by use of analytic relations. This is particularly the case when the relation involves a square root, as do many in table VI.20.

The least-square series approximation technique has the key advantage of not requiring much mathematical knowledge on the part of the user. It can be applied to approximate any function over a given interval, though the accuracy of the fit sometimes depends on the ingenuity of the user. For example, how should we deal with singularities, such as infinities? Two general approaches are possible. First, if the function has no zeroes in the interval, it might be better to approximate the inverse of the function. That is, if the function $f(x) = 1/(1 - x)$ is to be approximated, it would be better to fit $1/f(x)$, which is simply $(1 - x)$. If the pattern of singularities (ie: pole/zero structure in electrical engineering terms) is more complicated, then the variable to be used in the series may also need to be more complicated, such as $x/(x - a)$ instead of x. The series summation might then look like:

$$f(x) = \sum_{n=0}^{N} a(n) \left(\frac{x}{x-a}\right)^n \tag{VI.7}$$

The form given in this example might be tried for a case in which the function to be approximated goes to zero at $x = 0$ and to infinity at $x = a$. Series approximation is often an art.

$$
\begin{aligned}
\cos(x) &= \pm\sqrt{1 - \sin^2(x)} = \sin(\pi/2 - x) \\
\tan(x) &= \sin(x)/\cos(x) \\
\arccos(x) &= \pi/2 - \arctan(x/\sqrt{1 - x^2}) \\
\arcsin(x) &= \text{arcctan}(x/\sqrt{1 - x^2}) \\
\sinh(x) &= \tfrac{1}{2}(e^x - e^{-x}) \\
\cosh(x) &= \tfrac{1}{2}(e^x + e^{-x}) \\
\sinh(x) &= \tanh(x)/\sqrt{1 - \tanh^2(x)} \\
\cosh(x) &= \sqrt{1 + \sinh^2(x)} = 1/\sqrt{1 - \tanh^2(x)} \\
\tanh(x) &= \sinh(x)/\cosh(x) \\
\text{arcsinh}(x) &= \log_e(x + \sqrt{x^2 + 1}) \\
\text{arccosh}(x) &= \log_e(x \pm \sqrt{x^2 - 1}), \ (x \geq 1) + \text{sign used for principal value} \\
\text{arctanh}(x) &= \tfrac{1}{2}\log_e\left(\tfrac{1+x}{1-x}\right), \ (x^2 < 1) \\
e^x &= [1 + \tanh(x/2)]/[1 - \tanh(x/2)] \\
\arctan(x) &= \pi/2 - \arctan(1/x)
\end{aligned}
$$

Table VI.20 *Constitutive relationships.*

APPENDIX IA
SOFTWARE INDEX (by number)

APPENDIX IB
FUNCTION INDEX

APPENDIX IA   *SOFTWARE INDEX (by number)*

40000   Plotting subroutine for equally spaced data. Requires subroutine 40200.

40100   Plotting subroutine for coordinate pairs of data. Automatically sets spacing and scale. Requires subroutine 40200.

40200   Axis-printing subroutine. For use with subroutines 40000 and 40100.

40300   Rectangular complex-number addition subroutine.

40350   Rectangular complex-number subtraction subroutine.

40400   Rectangular (Cartesian)-to-polar coordinate-conversion subroutine. Requires the ATN, SQR (or SQRT), and exponentiation functions.

40450   Polar-to-rectangular coordinate-conversion subroutine. Requires the SIN and COS functions.

40500   Polar complex-number multiplication subroutine.

40550   Polar complex-number division subroutine.

40600   Rectangular complex-number multiplication subroutine. Requires subroutines 40400, 40450, and 40500.

40800   Rectangular complex-number division subroutine. Requires subroutines 40400, 40450, and 40550.

41100   Polar complex-number power subroutine. Requires the INT and exponentiation functions.

41150   Polar complex-number first root subroutine. Requires the exponentiation function.

41200 Rectangular complex-number exponentiation subroutine. Requires subroutines 40400, 40450, and 41100.

41300 Rectangular complex-number first root subroutine. Requires subroutines 40400, 40450, and 41150.

41400 Spherical-to-rectangular (Cartesian) coordinate conversion subroutine. Requires SIN and COS functions.

41450 Rectangular (Cartesian)-to-spherical coordinate conversion subroutine. Requires subroutine 40400.

41500 Vector addition subroutine.

41550 Vector subtraction subroutine.

41600 Vector dot product subroutine.

41650 Vector cross product subroutine.

41700 Vector length subroutine.

41750 Vector angle subroutine. Requires subroutines 41600 and 41850, as well as the square root function.

41800 Matrix addition subroutine.

41850 Matrix subtraction subroutine.

41900 Matrix multiplication subroutine.

41950 Matrix transpose subroutine.

42000 Diagonal matrix-generation subroutine.

42050 Matrix save (A in B) subroutine.

42075 Matrix save (B in A) subroutine.

42100 Matrix save (C in B) subroutine.

42125 Matrix save (B in C) subroutine.

42140 Matrix scalar multiplication subroutine.

42160 Matrix (A) clear subroutine.

42200 Matrix row-switching subroutine.

42250 Matrix row-manipulation subroutine (multiply row N1 by B and add to row N2).

42300 Matrix cofactor subroutine.

42350 Matrix determinant subroutine. Orders 1 thru 4. Requires subroutines 42050, 42075, 42125, and 42300.

42400 Matrix inversion subroutine.

42700 Eigenvalue subroutine. Uses the power method. Requires subroutine 41900.

42800 Matrix exponentiation subroutine. Requires subroutines 41900, 42000, 42100, 42125, and 42200.

42900 Linear pseudorandom number generator subroutine; empirical.

42925 Normal distribution pseudorandom number generator subroutine. Based on the RND function.

42950 Poisson distribution pseudorandom number generator subroutine. Based on the RND function.

43000 Exponential distribution pseudorandom number generator subroutine. Based on the RND function.

43025 Fermi distribution pseudorandom number generator subroutine. Based on the RND function.

43050 Cauchy distribution pseudorandom number generator subroutine. Based on the RND function.

43075 Gamma distribution pseudorandom number generator subroutine. Based on the RND function.

43100 Beta distribution pseudorandom number generator subroutine. Based on the RND function.

43150 Weibull distribution pseudorandom number generator subroutine. Based on the RND function.

43200   Series-summation subroutine. Basic subroutine called by other programs to perform standard summation.

43210   Sine series-approximation subroutine. Requires subroutines 43200 and 43300.

43225   Cosine series-approximation subroutine. Requires subroutines 43200 and 43310.

43245   Arctangent series-approximation subroutine. Requires subroutines 43200 and 43320.

43260   Logarithm to the base-10 series-approximation subroutine. Requires subroutines 43200 and 43340.

43280   Natural logarithm series-approximation subroutine. Requires subroutines 43200 and 43360.

43300   Sine series coefficients.

43310   Cosine series coefficients.

43320   Arctangent series coefficients.

43340   Logarithm to the base-10 series coefficients.

43360   Natural logarithm series coefficients.

43380   Exponential series coefficients.

43400   Power-of-ten series coefficients.

43450   Power-of-ten series-approximation subroutine. Requires subroutines 43200 and 43400.

43470   Exponent series-approximation subroutine. Requires subroutines 43200 and 43380.

APPENDIX IB *FUNCTION INDEX*

This appendix lists the subroutines presented in Volume I cross-indexed according to function. Program name mnemonics are also given to aid in program identification.

Complex Numbers

    Addition: Rectangular coordinate addition (ZADD); 40300.

    Division: Polar coordinate division (ZPOLDIV); 40550.
                    Rectangular coordinate division (ZRECTDIV); 40800.

    Multiplication: Polar coordinate multiplication (ZPOLMLT); 40500.
                          Rectangular coordinate multiplication (ZRECTMLT); 40600.

    Exponentiation: Polar coordinate exponentiation, Z^N (ZPOLPOW); 41100.
                          Rectangular coordinate exponentiation (ZRECTPOW); 41200.

    Root: Polar coordinate root, $Z^{1/N}$ (ZPOLRT); 41150.
            Rectangular coordinate root (ZRECTRT); 41300.

    Subtraction: Rectangular coordinate subtraction (ZRECTSUB); 40350.

Coordinate Conversion

    Polar-to-rectangular (POL/RECT); 40450.

    Rectangular-to-polar (RECT/POL); 40400.

    Rectangular-to-spherical (RECT/SPR); 41450.

    Spherical-to-rectangular (SPR/RECT); 41400.

Matrix Operations

Addition (MATADD); 41800.

Clear: Set all elements in A to zero (MATCLRA); 42225.

Cofactor: Determine Kth cofactor of A (MATCOFAT); 42300.

Determinant: Calculate determinant of A (MATDET); 42350.

Diagonal: Create diagonal matrix (MATDIAG); 42000.

Eigenvalues: Eigenvalue determination by the power method (EIGENPOW); 42700.

Exponentiate: Exp(Ax) (MATEXP); 42800.

Inversion (MATINV); 42400.

Multiplication (MATMULT); 41900.

Row add: Add multiple of one row to another (MATRADD); 42275.

Row shift (MATSWCH); 42250.

Save: Save source matrix in destination matrix (MATSAV <source> <destination>)

    MATSAVAB; 42050
    MATSAVBA; 42075
    MATSAVCB; 42100
    MATSAVBC; 42125
    MATSAVAC; 42150
    MATSAVCA; 42175

Scale: Multiply all elements by a constant (MATSCALE); 42200.

Subtraction (MATSUB); 41850.

Transpose (MATTRANS); 41950.

Plotting

Axis plot: Y axis plot (AXISPLOT); 40200.

Equally spaced data plot (EQAPLOT); 40000.

Two-dimensional data plot (DATAPLOT); 40100.

Random Number Generators

Beta distribution (BETA); 43100.

Binomial distribution (BINOMIAL); 42975.

Cauchy distribution (CAUCHY); 43050.

Exponential distribution (EXPONENT); 43000.

Fermi distribution (FERMI); 43025.

Gamma distribution (GAMMA); 43075.

Normal (shifted Gaussian) distribution (NORMAL); 42925.

Poisson distribution (POISSON); 42950.

Uniform distribution (LINEAR); 42900.

Weibull distribution (WEIBULL); 43150.

Series Approximations

Data subroutines:

Arctangent (ARCDATA); 43320.

Cosine (COSDATA); 43310.

Exponent, e^x (EXPDATA); 43380.

Logarithm, base 10 (LOGDATA); 43340.

Logarithm, base e (LNDATA); 43360.

Sine (SINDATA); 43300.

Ten to a power, 10^x (TENDATA), 43400.

Series summations:

    Arctangent (ARCTAN); 43245.

    Cosine (COSINE); 43225.

    Exponent, e^x (EXP(X)); **43470**

    Logarithm, base e (LN(X)); 43280.

    Logarithm, base 10 (LOG(X)); 43260.

    Series summation: general subroutine (SERSUM); 43200.

    Sine (SINE); 43210.

    Ten to a power, 10^x (TENPOW); 43450.

Vector Operations

    Addition (VECTADD); 41500.

    Angle between two vectors (VECTANGL); 41750.

    Cross product (VECTCURL); 41650.

    Dot product (VECTDOT); 41600.

    Length of a vector (VECTLEN); 41700.

    Subtraction (VECTSUB); 41550.

APPENDIX IIA

*Full Listings of Demonstration and
    Subroutine Programs*

APPENDIX A

APPENDIX IIA *Full Listings of Demonstration and Subroutine Programs*

The demonstration programs given in the text were concatenated in order of appearance and are reproduced in the next several pages. They are followed by the subroutine library.

Total demonstration program set length: 12337 bytes.

Total subroutine library length (uncompacted):   20836 bytes.

```
100 PRINT"PROGRAM FOR PLOTTING EQUALLY"
101 PRINT"SPACED DATA VALUES"
102 PRINT
103 PRINT"THE USER INPUTS THE STARTING "
104 PRINT"AND ENDING COORDINATES, ALONG "
105 PRINT"WITH THE NUMBER OF EVALUATION POINTS."
106 PRINT"THE PROGRAM WILL THEN PLOT "
107 PRINT"THE DATA"
108 PRINT
109 PRINT
110 REM INITIALIZATION
111 PRINT "INPUT TERMINAL WIDTH: ",
112 INPUT L
113 PRINT "BEGINNING CORDINATE: ",
114 INPUT X1
115 PRINT "ENDING COORDINATE: ",
116 INPUT X2
117 PRINT "NUMBER OF DATA POINTS: ",
118 INPUT N
119 DIM D(N+1)
120 REM DATA INPUT
121 PRINT"INPUT DATA"
122 FOR I=1 TO N
123 PRINT I,
124 INPUT D(I)
125 NEXT I
126 REM GO TO PLOTTING SUBROUTINE PROPER
127 GOSUB 40000
128 END
```

```
129 PRINT"PROGRAM FOR PLOTTING EQUALLY"
130 PRINT"SPACED FUNCTION VALUES"
131 PRINT
132 PRINT"THE USER INPUTS THE STARTING "
133 PRINT"AND ENDING COORDINATES, ALONG "
134 PRINT"WITH THE NUMBER OF EVALUATION POINTS."
135 PRINT"THE PROGRAM WILL THEN PLOT "
136 PRINT"THE FUNCTION OVER THAT RANGE."
137 PRINT
138 PRINT
139 REM INITIALIZATION
140 PRINT "INPUT TERMINAL WIDTH: ",
141 INPUT L
142 PRINT "BEGINNING CORDINATE: ",
143 INPUT X1
144 PRINT "ENDING COORDINATE: ",
145 INPUT X2
146 PRINT "NUMBER OF DATA POINTS: ",
147 INPUT N
148 DIM D(N+1)
149 REM FUNCTION EVALUATION
150 FOR I=1 TO N
151 REM *****INPUT FUNCTION BELOW*****
152 X=X1+(I-1)*(X2-X1)/(N-1)
153 D(I)=.1*X*X*X-3*X*X+2*X-3
154 NEXT I
155 REM GO TO PLOTTING SUBROUTINE PROPER
156 GOSUB 40000
157 END
158 PRINT"PROGRAM FOR PLOTTING COORDINATE"
159 PRINT"SETS OF DATA"
160 PRINT
161 PRINT"THE PROGRAM ASSUMES THAT THE DATA IS"
162 PRINT"SEQUENTIALLY ORDERED, FIRST BY ABSCISSA"
163 PRINT"VALUE, SECOND BY ORDINATE"
164 PRINT"THE PROGRAM WILL THEN PLOT "
165 PRINT"THE DATA."
166 PRINT
167 PRINT
168 REM INITIALIZATION
169 PRINT "INPUT TERMINAL WIDTH: ",
170 INPUT L
171 PRINT "NUMBER OF DATA POINTS: ",
172 INPUT N
173 DIM D(N+1),C(N+1),E(N+1)
174 REM DATA INPUT
175 PRINT "INPUT DATA IN ABSCISSA, ORDINATE PAIRS: "
176 FOR I=1 TO N
177 PRINT I,
```

```
178 INPUT C(I),D(I)
179 REM C(I)=ABSCISSA, D(I)=ORDINATE
180 NEXT I
181 REM GO TO PLOTTING SUBROUTINE PROPER
182 GOSUB 40100
183 END
184 REM PROGRAM TO DEMONSTRATE COMPLEX NUMBER
185 REM ADDITION, SUBTRACTION, MULTIPLICATION
186 REM AND DIVISION
187 PRINT"ENTER COMPLEX NUMBERS IN (X,Y) PAIRS:"
188 PRINT
189 PRINT"Z1= ",
190 INPUT X(1),Y(1)
191 PRINT
192 PRINT"Z2= ",
193 INPUT X(2),Y(2)
194 REM COMPLEX NUMBER ADDITION
195 GOSUB 40300
196 PRINT
197 PRINT
198 PRINT "Z1+Z2= ",X(3),
199 IF Y(3)>=0 THEN PRINT"+",
200 PRINT Y(3)," I"
201 PRINT
202 PRINT
203 REM COMPLEX NUMBER SUBTRACTION
204 GOSUB 40350
205 PRINT "Z1-Z2= ",X(3),
206 IF Y(3)>=0 THEN PRINT"+",
207 PRINT Y(3)," I"
208 PRINT
209 REM COMPLEX NUMBER MULTIPLICATION
210 GOSUB 40600
211 PRINT"Z1*Z2= ",X,
212 IF Y>=0 THEN PRINT" +",
213 PRINT Y," I"
214 PRINT
215 REM COMPLEX NUMBER DIVISION
216 GOSUB 40800
217 PRINT"Z1/Z2= "X,
218 IF Y>=0 THEN PRINT" +",
219 PRINT Y," I"
220 END
221 REM PROGRAM TO DEMONSTRATE
222 REM RECTANGULAR TO POLAR CONVERSION
223 PRINT"INPUT RECTANGULAR COORDINATES TO BE CONVERTED",
224 INPUT X,Y
225 REM CONVERSION
```

```
226 GOSUB 40400
227 PRINT
228 PRINT"POLAR COORDINATES ARE : RADIUS= ",U
229 PRINT"                              ANGLE=   ",V," RADIANS"
230 END
231 REM PROGRAM TO DEMONSTRATE
232 REM POLAR TO RECTANGULAR CONVERSION
233 PRINT"INPUT POLAR COORDINATES TO BE CONVERTED",
234 INPUT U,V
235 REM CONVERSION
236 GOSUB 40450
237 PRINT
238 PRINT"RECTANGULAR COORDINATES ARE (X,Y)= ",X," ,",Y
239 END
240 REM PROGRAM TO DEMONSTRATE
241 REM COMPLEX NUMBER TO A POWER
242 PRINT
243 PRINT"INPUT THE POWER DESIRED, FOLLOWED BY THE"
244 PRINT"COMPLEX NUMBER"
245 PRINT"N ",
246 INPUT N
247 PRINT"X ",
248 INPUT X
249 PRINT"Y ",
250 INPUT Y
251 REM POWER CALCULATION
252 GOSUB 41200
253 PRINT
254 PRINT"RESULT= ",X,
255 IF Y1>=0 THEN PRINT" +",
256 PRINT Y," I"
257 END
258 REM PROGRAM TO DEMONSTRATE
259 REM FINDING THE ROOTS OF A COMPLEX NUMBER
260 PRINT
261 PRINT"INPUT THE INTEGER ROOT DESIRED FOLLOWED"
262 PRINT"BY THE COMPLEX NUMBER"
263 PRINT"N ",
264 INPUT N
265 PRINT"X ",
266 REM X9 AND Y9 ARE STORED VALUES
267 INPUT X9
268 PRINT"Y ",
269 INPUT Y9
270 REM FIND N ROOTS
271 PRINT
272 PRINT
273 PRINT"ORDER    X+ YI"
274 PRINT"-----    ----------------"
```

```
275 PRINT
276 FOR M=1 TO N
277 REM CONVERT TO X,Y
278 X=X9
279 Y=Y9
280 GOSUB 41300
281 PRINT"   ",M,
282 PRINT"   ",INT(1000*X+.5)/1000,"   ",
283 IF Y>=0 THEN PRINT "+",
284 PRINT INT(1000*Y+.5)/1000," I"
285 NEXT M
286 END
287 REM PROGRAM TO DEMONSTRATE SPHERICAL TO
288 REM CARTESIAN COORDINATE CONVERSION
289 PRINT"INPUT RADIUS, U: ",
290 INPUT U
291 PRINT"INPUT ANGLE IN (X,Y) PLANE, V: ",
292 INPUT V
293 PRINT"INPUT ANGLE OFF Z AXIS, W: ",
294 INPUT W
295 GOSUB 41400
296 PRINT
297 PRINT"(X,Y,Z)= (",X," , ",Y," , ",Z,")"
298 END
299 REM PROGRAM TO DEMONSTRATE CARTESIAN TO
300 REM SPHERICAL COORDINATE CONVERSION
301 PRINT"INPUT X, Y AND Z: ",
302 INPUT X,Y,Z
303 GOSUB 41450
304 PRINT
305 PRINT"RADIUS= ",U
306 PRINT"ANGLE IN (X,Y) PLANE= ",V," RADIANS"
307 PRINT"ANGLE OFF Z AXIS= ",W," RADIANS"
308 END
309 REM PROGRAM TO DEMONSTRATE VECTOR SUBROUTINES
310 N=3
311 DIM A(N), B(N), C(N)
312 PRINT "INPUT VECTOR A (X,Y,Z): ",
313 INPUT A(1),A(2),A(3)
314 PRINT "INPUT VECTOR B (X,Y,Z): ",
315 INPUT B(1),B(2),B(3)
316 REM SUM
317 GOSUB 41500
318 PRINT
319 PRINT "A+B= (",C(1),", ",C(2),", ",C(3),")"
320 REM DIFFERENCE
321 GOSUB 41550
322 PRINT
```

```
323 PRINT"A-B= (",C(1),", ",C(2),", ",C(3),")"
324 REM CROSS PRODUCT
325 GOSUB 41650
326 PRINT
327 PRINT"AXB= (",C(1),", ",C(2),", ",C(3),")"
328 REM DOT PRODUCT
329 GOSUB 41600
330 PRINT
331 PRINT "A.B= ",C
332 REM VECTOR ANGLE
333 GOSUB 41750
334 PRINT
335 PRINT "ANGLE BETWEEN VECTORS A AND B= ",A," RADIANS"
336 REM VECTOR LENGTH
337 GOSUB 41700
338 PRINT
339 PRINT "LENGTH OF VECTOR B= ",L
340 PRINT
341 END
342 REM PROGRAM TO DEMONSTRATE MATRIX OPERATIONS
343 PRINT"INPUT THE ROW AND COLUMN DIMENSIONS OF A(I,J) AND B(I,J):"
344 PRINT"ROW SIZE= ",
345 INPUT M
346 PRINT"COLUMN SIZE= ",
347 INPUT N
348 REM FIND MAXIMUM MATRIX DIMENSION NEEDED
349 IF M>N THEN K=M+1
350 IF N>=M THEN K=N+1
351 DIM A(K,K), B(K,K), C(K,K)
352 REM INPUT MATRIX A
353 PRINT
354 PRINT"INPUT MATRIX A ROW BY ROW"
355 PRINT
356 FOR I=1 TO M
357 PRINT"INPUT THE ",N," ELEMENTS OF ROW ",I," :"
358 FOR J=1 TO N
359 INPUT A(I,J)
360 NEXT J
361 PRINT
362 NEXT I
363 PRINT
364 REM PRINT MATRIX A
365 REM MOVE A TO C
366 N1=M
367 N2=N
368 N3=0
369 GOSUB 42150
370 PRINT"MATRIX A="
```

```
371 PRINT
372 REM PRINTING SUBROUTINE
373 GOSUB 1020
374 PRINT
375 REM INPUT MATRIX B
376 PRINT
377 PRINT"INPUT MATRIX B ROW BY ROW"
378 PRINT
379 FOR I=1 TO M
380 PRINT"INPUT THE ",N," ELEMENTS OF ROW ",I," :"
381 FOR J=1 TO N
382 INPUT B(I,J)
383 NEXT J
384 PRINT
385 NEXT I
386 PRINT
387 REM MOVE B TO C
388 GOSUB 42125
389 PRINT
390 PRINT"MATRIX B="
391 PRINT
392 REM PRINTING SUBROUTINE
393 GOSUB 1020
394 PRINT
395 REM MATRIX ADDITION
396 GOSUB 41800
397 PRINT"A+B="
398 REM PRINTING SUBROUTINE
399 GOSUB 1020
400 REM MATRIX SUBTRACTION
401 GOSUB 41850
402 PRINT"A-B="
403 REM PRINTING SUBROUTINE
404 GOSUB 1020
405 REM MATRIX TRANSPOSE (A)
406 REM FIRST, SAVE B IN C
407 GOSUB 42125
408 REM PUT TRANSPOSE OF A IN B
409 GOSUB 41950
410 REM MOVE TRANSPOSE OF A TO A
411 N1=N
412 N2=M
413 GOSUB 42075
414 REM RETURN C TO B
415 N1=M
416 N2=N
417 GOSUB 42100
418 PRINT
419 PRINT"A(TRANSPOSE)="
```

```
420 REM SWITCH ROW AND COLUMN SIZES FOR THE TRANSPOSE PRINT
421 N2=M
422 N1=N
423 M=N1
424 N=N2
425 REM MOVE A TO C FOR PRINTING
426 GOSUB 42150
427 PRINT
428 REM PRINTING SUBROUTINE
429 GOSUB 1020
430 REM DETERMINE PRODUCT OF A(TRANSPOSE) AND B
431 M1=M
432 N1=N
433 M2=N
434 N2=M
435 REM MATRIX MULTIPLICATION SUBROUTINE
436 GOSUB 41900
437 PRINT"A(TRANSPOSE) X B="
438 PRINT
439 REM PRINTING SUBROUTINE
440 N=M
441 GOSUB 1020
442 PRINT
443 END
444 REM PROGRAM TO DEMONSTRATE SPECIAL MATRIX OPERATIONS
445 REM FIRST CREATE A DIAGONAL MATRIX
446 PRINT"WHAT SIZE MATRIX IS DESIRED",
447 INPUT N
448 PRINT
449 PRINT"THIS MATRIX WILL HAVE ONLY DIAGONAL ELEMENTS.",
450 PRINT" WHAT VALUE SHOULD THEY BE",
451 INPUT B
452 DIM A(N+1,N+1),B(N+1,N+1),C(N+1,N+1)
453 REM GO TO DIAGONAL MATRIX CREATION SUBROUTINE
454 GOSUB 42000
455 REM SAVE MATRIX IN BOTH A AND C
456 N1=N
457 N2=N
458 N3=0
459 GOSUB 42075
460 GOSUB 42125
461 REM PRINT RESULT
462 M=N
463 GOSUB 1020
464 REM SWITCH TWO ROWS
465 PRINT
466 PRINT"INPUT THE NUMBERS OF THE TWO ROWS TO BE SWITCHED",
467 INPUT N1,N2
468 REM GO TO SWITCHING SUBROUTINE
```

```
469 GOSUB 42250
470 REM MOVE RESULT TO C AND PRINT
471 N1=N
472 N2=N
473 GOSUB 42150
474 PRINT
475 GOSUB 1020
476 PRINT"INPUT SCALE FACTOR TO BE MULTIPLIED BY: ",
477 INPUT B
478 GOSUB 42200
479 REM PRINT RESULT
480 GOSUB 42150
481 GOSUB 1020
482 PRINT
483 PRINT"INPUT WHAT FRACTION OF ROW N1 IS TO BE ADDED TO ROW N2"
484 PRINT"FRACTION: ",
485 INPUT B
486 PRINT"N1: ",
487 INPUT N1
488 PRINT"N2: ",
489 INPUT N2
490 REM GO TO OPERATION
491 GOSUB 42275
492 REM PRINT RESULT
493 PRINT
494 N2=N
495 N1=N
496 GOSUB 42150
497 GOSUB 1020
498 PRINT"CLEAR MATRIX"
499 N1=N
500 N2=N
501 GOSUB 42225
502 REM PRINT RESULT
503 PRINT
504 GOSUB 42150
505 GOSUB 1020
506 END
507 REM PROGRAM TO DEMONSTRATE DETERMINANT
508 PRINT"INPUT MATRIX SIZE: ",
509 INPUT N
510 DIM A(N+1,N+1),B(N+1,N+1),C(N+1,N+1)
511 PRINT
512 REM INPUT MATRIX
513 FOR I=1 TO N
514 PRINT"INPUT ROW ",I
515 FOR J=1 TO N
516 INPUT A(I,J)
517 NEXT J
```

```
518 PRINT
519 NEXT I
520 PRINT
521 PRINT "MATRIX A="
522 REM MOVE A TO C FOR PRINTING
523 N1=N
524 N2=N
525 N3=0
526 GOSUB 42150
527 REM PRINT
528 M=N
529 GOSUB 1020
530 PRINT
531 REM EVALUATE DETERMINANT
532 GOSUB 42350
533 PRINT"DETERMINANT= ",D
534 END
535 REM PROGRAM TO DEMONSTRATE MATRIX INVERSION
536 PRINT"INPUT DIMENSION OF MATRIX TO BE INVERTED: ",
537 INPUT N
538 DIM A(N+1,N+1),B(N+1,2*N+2),C(N+1,N+1)
539 PRINT
540 FOR I=1 TO N
541 PRINT"INPUT ROW ",I," :"
542 FOR J=1 TO N
543 INPUT A(I,J)
544 NEXT J
545 PRINT
546 NEXT I
547 PRINT
548 PRINT"MATRIX A="
549 N1=N
550 N2=N
551 N3=0
552 REM MOVE A TO C
553 GOSUB 42150
554 M=N
555 GOSUB 1020
556 REM GOTO INVERSION SUBROUTINE
557 GOSUB 42400
558 REM MOVE B TO C TO PRINT
559 PRINT"INVERSE OF A="
560 PRINT
561 GOSUB 42125
562 M=N
563 GOSUB 1020
564 REM CHECK RESULTS
565 REM MULTIPLY A TIMES B
```

```
566 M1=N
567 M2=N
568 GOSUB 41900
569 M=N
570 REM PRINT RESULTS
571 PRINT
572 PRINT"MATRIX A TIMES INVERSE MATRIX A="
573 GOSUB 1020
574 END
575 REM PROGRAM TO DEMONSTRATE USE OF INVERSE TO SOLVE
576 REM SIMULTANEOUS EQUATIONS
577 PRINT"INPUT SIZE OF EQUATION (NUMBER OF UNKNOWNS): ",
578 INPUT N
579 DIM A(N+1,N+1),B(N+1,2*N+2),C(N+1,N+1)
580 PRINT
581 FOR I=1 TO N
582 PRINT"INPUT ROW ",I," OF COEFFICIENT MATRIX"
583 FOR J=1 TO N
584 INPUT A(I,J)
585 NEXT J
586 PRINT
587 PRINT"INPUT CONSTANT WHICH ROW EQUATION EQUALS: ",
588 INPUT B(I,1)
589 PRINT
590 NEXT I
591 PRINT
592 PRINT"MATRIX A="
594 N1=N
595 N2=N
596 N3=0
597 REM MOVE A TO C
598 GOSUB 42150
599 REM PRINT A
600 M=N
602 GOSUB 1020
603 REM MOVE CONSTANT VECTOR TO C
604 GOSUB 42125
605 REM PRINT CONSTANT VECTOR
606 PRINT
607 PRINT"CONSTANT VECTOR ="
608 N=1
609 GOSUB 1020
610 N=M
611 REM OBTAIN INVERSE
612 GOSUB 42400
613 REM MOVE RESULT IN B TO A
614 GOSUB 42075
615 REM MOVE CONSTANT VECTOR IN C TO B
616 GOSUB 42100
```

```
617 REM MULTIPLY INVERSE TIMES CONSTANT VECTOR
618 M1=N
619 N1=N
620 M2=N
621 N2=1
622 GOSUB 41900
623 REM RESULT IS IN C. PRINT C
624 M=N
625 N=1
626 PRINT"SOLUTION VECTOR="
627 GOSUB 1020
628 END
629 REM PROGRAM TO DEMONSTRATE OBTAINING THE LARGEST EIGENVALUE OF A MATRIX
630 PRINT"WHAT IS THE SIZE OF THE MATRIX: ",
631 INPUT N
632 PRINT
633 DIM A(N+1,N+1),B(N+1,N+1),C(N+1,N+1)
634 FOR I=1 TO N
635 PRINT"INPUT ROW ",I
636 FOR J=1 TO N
637 INPUT A(I,J)
638 NEXT J
639 PRINT
640 NEXT I
641 PRINT
642 PRINT"INPUT LEVEL OF ACCURACY: ",
643 INPUT E
644 PRINT"INPUT NUMBER OF ITERATIONS AT WHICH TO STOP: ",
645 INPUT D1
646 REM FIND EIGENVALUE
647 GOSUB 42700
648 PRINT
649 PRINT"EIGENVALUE= ",A
650 PRINT
651 PRINT"EIGENVECTOR="
652 FOR I=1 TO N
653 PRINT B(I,1)
654 NEXT I
655 PRINT
656 PRINT"NUMBER OF ITERATIONS= ",D
657 END
658 REM PROGRAM TO DEMONSTRATE THE MATRIX EXPONENT SUBROUTINE
659 PRINT"INPUT THE SIZE OF THE MATRIX TO BE EXPONENTIATED",
660 INPUT N
661 DIM A(N+1,N+1), B(N+1,N+1), C(N+1,N+1), D(N+1,N+1)
662 FOR I=1 TO N
663 PRINT"INPUT ROW ",I
664 FOR J=1 TO N
665 INPUT A(I,J)
```

```
666 NEXT J
667 PRINT
668 NEXT I
669 PRINT
670 REM PRINT MATRIX A
671 N1=N
672 N2=N
673 N3=0
674 M=N
675 PRINT"MATRIX A="
676 PRINT
677 REM MOVE A TO C
678 GOSUB 42150
679 REM PRINT
680 GOSUB 1020
681 PRINT"INPUT THE VARIABLE X",
682 INPUT X
683 PRINT
684 PRINT"INPUT THE NUMBER OF TERMS TO BE CALCULATED",
685 INPUT K2
686 REM OBTAIN EXPONENT
687 GOSUB 42800
688 PRINT
689 PRINT"EXP(AX)= "
690 PRINT
691 GOSUB 1020
692 PRINT
693 END
694 REM PROGRAM TO DEMONSTRATE THE USE OF THE UNIFORM
695 REM RANDOM NUMBER GENERATOR, RND(0)
696 L=80
697 REM SET PLOT WIDTH
698 N=21
699 DIM C(22),D(22),E(22)
700 FOR I=1 TO 21
701 D(I)=0
702 C(I)=(I-1)/10
703 NEXT I
704 REM SET SEED
705 Z2=RND(.5)
706 FOR I2=1 TO 1000
707 Z2=10*RND(0)+1
708 Z2=INT(Z2)
709 D(Z2)=D(Z2)+1
710 NEXT I2
711 PRINT"DISTRIBUTION OF THE 1000 RANDOM NUMBERS"
712 PRINT"IN TERMS OF THE NUMBER PER INTERVAL OF LENGTH 1/10"
713 PRINT
714 PRINT
```

```
715 GOSUB 40100
716 PRINT
717 PRINT
718 END
719 REM SINE AND COSINE APPROXIMATION DEMONSTRATION
720 PRINT"   X          SIN(X)              DELTA            COS(X)            DELTA"
721 PRINT"  ---        --------            -------          -------          -------"
722 PRINT
723 FOR X=-5 TO 5 STEP .2
724 GOSUB 43210
725 PRINT %4F1,X,%19F14,Y,%13E2,(SIN(X)-Y),
726 GOSUB 43225
727 PRINT %26F14,Y,%13E2,(COS(X)-Y)
728 NEXT X
729 END
730 REM ARCTANGENT APPROXIMATION DEMONSTRATION
731 PRINT"   X        ARCTANGENT            DELTA"
732 PRINT"  ---      ------------          -------"
733 PRINT
734 FOR X=-5 TO 5 STEP .2
735 GOSUB 43245
736 PRINT %4F1,X,%23F14,Y,%14E2,(ATN(X)-Y)
737 NEXT X
738 END
739 REM NATURAL LOGARITHM AND EXPONENT APPROXIMATION DEMONSTRATION
740 PRINT "    X         LN(X)              DELTA          EXP(LN(X))         DELTA"
741 PRINT "   ---       -------            -------        -------------      -------"
742 FOR X=.1 TO 5 STEP .1
743 GOSUB 43280
744 PRINT %3F1," ",X,%19F14,Y,%13E2,(LOG(X)-Y),
745 F=X
746 X=Y
747 GOSUB 43470
748 PRINT %25F14,Y,%14E2,(F-Y)
749 X=F
750 NEXT X
751 END
752 REM BASE 10 LOGARITHM AND EXPONENT APPROXIMATION DEMONSTRATION
753 PRINT "   X         LOG(X)              DELTA           10^(LOG(X))        DELTA"
754 PRINT "  ---       --------            -------         -------------     --------"
755 FOR X=.1 TO 5 STEP .1
756 GOSUB 43260
757 PRINT %3F1," ",X,%19F14,Y,%13E2,(.434294481903251*LOG(X)-Y),
758 F=X
759 X=Y
760 GOSUB 43450
761 PRINT %25F14,Y,%14E2,(F-Y)
762 X=F
763 NEXT X
764 END
1000 REM ********************
1010 REM PRINTING SUBROUTINE
1020 PRINT
1030 FOR I=1 TO M
1040 FOR J=1 TO N
1050 PRINT TAB(6*J),INT(C(I,J)*100+.5)/100,
1060 NEXT J
1070 PRINT
1080 NEXT I
1090 PRINT
1100 RETURN
```

APPENDIX IIA

Listing of North Star Demonstration Programs Vol. 1

```
39998 REM PLOTTING SUBROUTINE (EQAPLOT)
39999 REM SHIFT DATA TO NON-NEGATIVE
40000 B=100000000
40001 REM FIND MINIMUM DATA VALUE
40002 FOR I=1 TO N
40003 IF B>D(I) THEN B=D(I)
40004 NEXT I
40005 REM SUBTRACT MINIMUM VALUE FROM ALL DATA
40006 FOR I=1 TO N
40007 D(I)=D(I)-B
40008 NEXT I
40009 REM FIND MAX. SHIFTED DATA VALUE
40010 C=0
40011 FOR I=1 TO N
40012 IF C<D(I) THEN C=D(I)
40013 NEXT I
40014 REM DETERMINE PRINTING SCALE VALUE
40015 A=L/C
40016 REM FIND TAB POSITION OF ZERO
40017 E=A*ABS(B)
40018 PRINT
40019 PRINT
40020 PRINT"***** DATA PLOT (SCALED) *****"
40021 PRINT
40022 PRINT
40023 PRINT"MIN. ORDINATE= ",B,"   MAX. ORDINATE= ",C+B
40024 PRINT"INITIAL ABSCISSA VALUE= ",X1
40025 PRINT
40026 PRINT
40027 REM IF B IS POSITIVE, SKIP ZERO LABEL
40028 IF B>0 THEN GOTO 40034
40029 REM IF DATA ARE ALL BELOW ZERO, SKIP LABEL
40030 IF ABS(B)>C THEN GOTO 40034
40031 REM LABEL ZERO
40032 PRINT TAB(E),"0"
40033 REM GO TO AXIS PRINT SUBROUTINE
40034 GOSUB 40200
40035 FOR I=1 TO N
40036 REM INSERT LINE FEED FOR AUTO SPACING
40037 FOR K=1 TO (INT(0.6*L/N))
40038 PRINT":",TAB(L),":"
40039 NEXT K
40040 REM LOCATE DATUM POSITION
40041 E2=A*D(I)
40042 REM FORMATTED PRINT
40043 IF E2>=1 THEN GOTO 40046
40044 PRINT"*",
40045 GOTO 40049
40046 PRINT":",
```

```
40047 PRINT TAB(E2),"*",
40048 IF INT(E2)=L THEN GOTO 40050
40049 PRINT TAB(L),":",
40050 PRINT
40051 NEXT I
40052 REM GO TO AXIS PRINT SUBROUTINE
40053 GOSUB 40200
40054 PRINT
40055 PRINT
40056 PRINT"END ABSCISSA VALUE= ",X2
40057 PRINT
40058 PRINT
40059 REM RETURN TO DATA SOURCE PROGRAM
40060 RETURN
40098 REM TWO DIMENSIONAL DATA PLOTTING SUBROUTINE (DATAPLOT)
40099 REM SHIFT DATA TO NON-NEGATIVE
40100 B=100000000
40101 REM FIND MINIMUM DATA VALUE
40102 FOR I=1 TO N
40103 IF B>D(I) THEN B=D(I)
40104 NEXT I
40105 REM SUBTRACT MINIMUM VALUE FROM ALL DATA
40106 FOR I=1 TO N
40107 D(I)=D(I)-B
40108 NEXT I
40109 REM FIND MAX. SHIFTED DATA VALUE
40110 C=0
40111 FOR I=1 TO N
40112 IF C<D(I) THEN C=D(I)
40113 NEXT I
40114 REM DETERMINE E(I), THE ABSCISSA SPACINGS
40115 E(0)=0
40116 E(N)=1
40117 FOR I=2 TO N
40118 E(I-1)=INT(0.5*(C(I)-C(I-1))*L/(C(N)-C(1))+.5)
40119 REM SPACING SCALED ACCORDING TO LINE WIDTH
40120 NEXT I
40121 REM DETERMINE PRINTING SCALE VALUE
40122 A=L/C
40123 REM FIND TAB POSITION OF ZERO
40124 E=A*ABS(B)
40125 PRINT
40126 PRINT
40127 PRINT"***** DATA PLOT (SCALED) *****"
40128 PRINT
40129 PRINT
40130 PRINT"MIN. ORDINATE=  ",B,"    MAX. ORDINATE= ",C+B
40131 PRINT"INITIAL ABSCISSA VALUE= ",C(1)
40132 PRINT
```

```
40133 PRINT
40134 REM IF B IS POSITIVE, SKIP ZERO LABEL
40135 IF B>0 THEN GOTO 40141
40136 REM IF DATA ARE ALL BELOW ZERO, SKIP LABEL
40137 IF ABS(B)>C THEN GOTO 40141
40138 REM LABEL ZERO
40139 PRINT TAB(E),"0"
40140 REM GO TO AXIS PRINT SUBROUTINE
40141 GOSUB 40200
40142 FOR I=1 TO N
40143 REM INSERT FEED FOR ABSCISSA SPACING
40144 FOR K=1 TO E(I-1)
40145 PRINT":",TAB(L),":"
40146 NEXT K
40147 REM LOCATE DATUM POSITION
40148 E2=A*D(I)
40149 REM TEST FOR MULTIPLE ORDINATE ABSCISSA
40150 IF E(I)=0 THEN GOTO 40161
40151 REM FORMATTED PRINT
40152 IF E2>=1 THEN GOTO 40155
40153 PRINT"*",
40154 GOTO 40158
40155 PRINT":",
40156 PRINT TAB(E2),"*",
40157 IF INT(E2)=L THEN GOTO 40159
40158 PRINT TAB(L),":",
40159 PRINT
40160 GOTO 40175
40161 REM TEST FOR OVERLAY OF POINTS
40162 REM IF DUPLICATE POINT, SKIP TO END
40163 IF D(I)=D(I+1) THEN GOTO 40175
40164 IF E2>=1 THEN GOTO 40168
40165 REM PRINT DOUBLE VALUES ON ONE LINE
40166 PRINT"*",TAB(A*D(I+1)),"*",
40167 GOTO 40171
40168 PRINT":",
40169 PRINT TAB(E2),"*",TAB(A*D(I+1)),"*",
40170 REM TEST IF LINE LIMIT REACHED
40171 IF INT(A*D(I+1))=L THEN GOTO 40173
40172 PRINT TAB(L),":",
40173 PRINT
40174 I=I+1
40175 NEXT I
40176 REM GO TO AXIS PRINT SUBROUTINE
40177 GOSUB 40200
40178 PRINT
40179 PRINT
40180 PRINT"END ABSCISSA VALUE= ",C(N)
40181 PRINT
```

```
40182 PRINT
40183 REM RETURN TO DATA SOURCE PROGRAM
40184 RETURN
40199 REM AXIS PLOT (AXISPLOT)
40200 E3=E-5*INT(E/5)
40201 REM IF B IS POSTIVE, THEN SKIP ZERO LABEL
40202 IF B>0 THEN E3=0
40203 REM IF B IS GREATER THAN THE LARGEST VALUE, SKIP
40204 IF ABS(B)>C THEN E3=0
40205 FOR J=1 TO E3
40206 PRINT"-",
40207 NEXT J
40208 FOR J=1 TO (L-E3)/5
40209 PRINT"I----",
40210 NEXT J
40211 PRINT"I",
40212 E4=(J-1)*5+1+E3
40213 IF E4=L+1 THEN PRINT
40214 IF E4=L+1 THEN GOTO 40221
40215 E4=E4+1
40216 IF E4>=L+1 THEN GOTO 40219
40217 PRINT"-",
40218 GOTO 40215
40219 PRINT":"
40220 REM RETURN TO MAIN PLOTTING PROGRAM
40221  RETURN
40299 REM COMPLEX NUMBER ADDITION SUBROUTINE (ZADD)
40300 X(3)=X(1)+X(2)
40301 Y(3)=Y(1)+Y(2)
40302 RETURN
40349 REM COMPLEX NUMBER SUBTRACTION SUBROUTINE (ZSUB)
40350 X(3)=X(1)-X(2)
40351 Y(3)=Y(1)-Y(2)
40352 RETURN
40399 REM RECTANGULAR TO POLAR CONVERSION SUBROUTINE (RECT/POL)
40400 U=SQRT(X*X+Y*Y)
40401 REM GUARD AGAINST AMBIGUOUS VECTOR
40402 IF Y=0 THEN Y=(.1)^30
40403 REM GUARD AGAINST DIVIDE BY ZERO
40404 IF X=0 THEN X=(.1)^30
40405 REM SOME BASICS REQUIRE A SIMPLE ARGUMENT
40406 W=Y/X
40407 V=ATN(W)
40408 REM CHECK QUADRANT AND ADJUST
40409 IF X<0 THEN V=V+3.1415926535
40410 IF V<0 THEN V=V+6.2831853072
40411 RETURN
40449 REM POLAR TO RECTANGULAR CONVERSION SUBROUTINE (POL/RECT)
40450 X=U*COS(V)
```

```
40451 Y=U*SIN(V)
40452 RETURN
40499 REM POLAR MULTIPLICATION SUBROUTINE (ZPOLMLT)
40500 U=U(1)*U(2)
40501 V=V(1)+V(2)
40502 IF V>=6.2831853072 THEN V=V-6.2831853072
40503 RETURN
40549 REM POLAR DIVISION SUBROUTINE (ZPOLDIV)
40550 U=U(1)/U(2)
40551 V=V(1)-V(2)
40552 IF V<0 THEN V=V+6.2831853072
40553 RETURN
40599 REM RECTANGULAR COMPLEX NUMBER MULTIPLICATION SUBROUTINE (ZRECTMLT)
40600 X=X(1)
40601 Y=Y(1)
40602 REM RECTANGULAR TO POLAR CONVERSION
40603 GOSUB 40400
40604 U(1)=U
40605 V(1)=V
40606 X=X(2)
40607 Y=Y(2)
40608 REM RECTANGULAR TO POLAR CONVERSION
40609 GOSUB 40400
40610 U(2)=U
40611 V(2)=V
40612 REM POLAR MULTIPLICATION
40613 GOSUB 40500
40614 REM POLAR TO RECTANGULAR CONVERSION
40615 GOSUB 40450
40616 RETURN
40799 REM RECTANGULAR COMPLEX NUMBER DIVISION SUBROUTINE (ZRECTDIV)
40800 X=X(1)
40801 Y=Y(1)
40802 REM RECTANGULAR TO POLAR CONVERSION
40803 GOSUB 40400
40804 U(1)=U
40805 V(1)=V
40806 X=X(2)
40807 Y=Y(2)
40808 REM RECTANGULAR TO POLAR CONVERSION
40809 GOSUB 40400
40810 U(2)=U
40811 V(2)=V
40812 REM POLAR COMPLEX NUMBER DIVISION
40813 GOSUB 40550
40814 REM POLAR TO RECTANGULAR CONVERSION
40815 GOSUB 40450
40816 RETURN
41099 REM POLAR POWER SUBROUTINE (ZPOLPOW)
41100 U1=U^N
41101 V1=N*V
41102 V1=V1-6.2831853072*INT(V1/6.2831853072)
41103 RETURN
41149 REM POLAR (FIRST) ROOT SUBROUTINE (ZPOLRT)
41150 U1=U^(1/N)
41151 V1=V/N
```

```
41152 RETURN
41198 REM RECTANGULAR COMPLEX NUMBER POWER SUBROUTINE (ZRECTPOW)
41199 REM RECTANGULAR TO POLAR CONVERSION
41200 GOSUB 40400
41201 REM POLAR POWER
41202 GOSUB 41100
41203 REM CHANGE VARIABLE FOR CONVERSION
41204 U=U1
41205 V=V1
41206 REM POLAR TO RECTANGULAR CONVERSION
41207 GOSUB 40450
41208 RETURN
41298 REM RECTANGULAR COMPLEX NUMBER ROOT SUBROUTINE (ZRECTRT)
41299 REM RECTANGULAR TO POLAR CONVERSION
41300 GOSUB 40400
41301 REM POLAR (FIRST) ROOT
41302 GOSUB 41150
41303 U=U1
41304 REM FIND M ORDER ROOT
41305 REM M=1 CORRESPONDS TO THE FIRST ROOT
41306 V=V1+6.2831853072*(M-1)/N
41307 REM POLAR TO RECTANGULAR CONVERSION
41308 GOSUB 40450
41309 RETURN
41399 REM SPHERICAL TO RECTANGULAR (CARTESIAN) CONVERSION SUBROUTINE (SPR/RECT)
41400 X=U*(SIN(W))*COS(V)
41401 Y=U*(SIN(W))*SIN(V)
41402 Z=U*COS(W)
41403 RETURN
41448 REM RECTANGULAR (CARTESIAN) TO SPHERICAL CONVERSION SUBROUTINE (RECT/SPR)
41449 REM RECTANGULAR TO POLAR CONVERSION
41450 GOSUB 40400
41451 REM SAVE AND CHANGE VARIABLES
41452 V1=V
41453 X=U
41454 Y=Z
41455 REM RECTANGULAR TO POLAR CONVERSION
41456 GOSUB 40400
41457 IF V>1.5707963268 THEN V=V-6.28318553072
41458 W=1.5707963268-V
41459 V=V1
41460 RETURN
41498 REM VECTOR ADDITION SUBROUTINE (VECTADD)
41499 REM C=A+B
41500 FOR I=1 TO N
41501 C(I)=A(I)+B(I)
41502 NEXT I
41503 RETURN
41548 REM VECTOR SUBTRACTION SUBROUTINE (VECTSUB)
41549 REM C=A-B
41550 FOR I=1 TO N
41551 C(I)=A(I)-B(I)
41552 NEXT I
41553 RETURN
41598 REM VECTOR DOT PRODUCT SUBROUTINE (VECTDOT)
41599 REM C=A.B
```

```
41600 C=0
41601 FOR I=1 TO N
41602 C=C+A(I)*B(I)
41603 NEXT I
41604 RETURN
41648 REM VECTOR CROSS PRODUCT SUBROUTINE (VECTCURL)
41649 REM C=A X B
41650 C(1)=A(2)*B(3)-A(3)*B(2)
41651 C(2)=A(3)*B(1)-A(1)*B(3)
41652 C(3)=A(1)*B(2)-A(2)*B(1)
41653 RETURN
41699 REM VECTOR LENGTH SUBROUTINE (VECTLEN)
41700 L=0
41701 FOR I=1 TO N
41702 L=L+A(I)*A(I)
41703 NEXT I
41704 L=SQRT(L)
41705 RETURN
41747 REM VECTOR ANGLE SUBROUTINE (VECTANGL)
41748 REM ANGLE BETWEEN A AND B
41749 REM FIND DOT PRODUCT
41750 GOSUB 41600
41751 REM FIND LENGTH OF A
41752 GOSUB 41700
41753 REM SAVE VALUE
41754 L1=L
41755 REM FIND LENGTH OF B
41756 FOR I=1 TO N
41757 A(I)=B(I)
41758 NEXT I
41759 GOSUB 41700
41760 E=C/(L*L1)+(.1)^30
41761 E=SQRT(1-E*E)/E
41762 A=ATN(E)
41763 IF C<0 THEN A=3.1415926536-A
41764 RETURN
41798 REM MATRIX ADDITION SUBROUTINE (MATADD)
41799 REM C=A+B
41800 FOR I=1 TO M
41801 FOR J=1 TO N
41802 C(I,J)=A(I,J)+B(I,J)
41803 NEXT J
41804 NEXT I
41805 RETURN
41848 REM MATRIX SUBTRACTION SUBROUTINE (MATSUB)
41849 REM C=A-B
41850 FOR I=1 TO M
41851 FOR J=1 TO N
41852 C(I,J)=A(I,J)-B(I,J)
```

```
41853 NEXT J
41854 NEXT I
41855 RETURN
41898 REM MATRIX MULTIPLICATION SUBROUTINE (MATMULT)
41899 REM C=A X B    A IS M1 BY N1    B IS M2 BY N2    C IS M1 BY N2
41900 FOR I=1 TO M1
41901 FOR J=1 TO N2
41902 C(I,J)=0
41903 FOR K=1 TO N1
41904 C(I,J)=C(I,J)+A(I,K)*B(K,J)
41905 NEXT K
41906 NEXT J
41907 NEXT I
41908 RETURN
41948 REM MATRIX TRANSPOSE SUBROUTINE (MATTRANS)
41949 REM B=TRANSPOSE(A)
41950 FOR I=1 TO N
41951 FOR J=1 TO M
41952 B(I,J)=A(J,I)
41953 NEXT J
41954 NEXT I
41955 RETURN
41998 REM DIAGONAL MATRIX CREATION SUBROUTINE (MATDIAG)
41999 REM MATRIX B(I,J) IS THE IDENTITY MATRIX TIMES B
42000 FOR I=1 TO N
42001 FOR J=1 TO N
42002 B(I,J)=0
42003 IF I=J THEN B(I,J)=B
42004 NEXT J
42005 NEXT I
42006 RETURN
42048 REM MATRIX SAVE (A IN B) SUBROUTINE (MATSAVAB)
42049 REM N1,N2 AND N3 ARE INPUT INDICES
42050 IF N1*N2*N3=0 THEN GOTO 42060
42051 REM CHECK DIMENSION
42052 FOR I1=1 TO N1
42053 FOR I2=1 TO N2
42054 FOR I3=1 TO N3
42055 B(I1,I2,I3)=A(I1,I2,I3)
42056 NEXT I3
42057 NEXT I2
42058 NEXT I1
42059 RETURN
42060 IF N1*N2=0 THEN GOTO 42067
42061 FOR I1=1 TO N1
42062 FOR I2=1 TO N2
42063 B(I1,I2)=A(I1,I2)
42064 NEXT I2
42065 NEXT I1
```

```
42066 RETURN
42067 IF N1=0 THEN RETURN
42068 FOR I1=1 TO N1
42069 B(I1)=A(I1)
42070 NEXT I1
42071 RETURN
42073 REM MATRIX SAVE (B IN A) SUBROUTINE (MATSAVBA)
42074 REM N1,N2 AND N3 ARE INPUT INDICES
42075 IF N1*N2*N3=0 THEN GOTO 42085
42076 REM CHECK DIMENSION
42077 FOR I1=1 TO N1
42078 FOR I2=1 TO N2
42079 FOR I3=1 TO N3
42080 A(I1,I2,I3)=B(I1,I2,I3)
42081 NEXT I3
42082 NEXT I2
42083 NEXT I1
42084 RETURN
42085 IF N1*N2=0 THEN GOTO 42092
42086 FOR I1=1 TO N1
42087 FOR I2=1 TO N2
42088 A(I1,I2)=B(I1,I2)
42089 NEXT I2
42090 NEXT I1
42091 RETURN
42092 IF N1=0 THEN RETURN
42093 FOR I1=1 TO N1
42094 A(I1)=B(I1)
42095 NEXT I1
42096 RETURN
42098 REM MATRIX SAVE (C IN B) SUBROUTINE (MATSAVCB)
42099 REM N1,N2 AND N3 ARE INPUT INDICES
42100 IF N1*N2*N3=0 THEN GOTO 42110
42101 REM CHECK DIMENSION
42102 FOR I1=1 TO N1
42103 FOR I2=1 TO N2
42104 FOR I3=1 TO N3
42105 B(I1,I2,I3)=C(I1,I2,I3)
42106 NEXT I3
42107 NEXT I2
42108 NEXT I1
42109 RETURN
42110 IF N1*N2=0 THEN GOTO 42117
42111 FOR I1=1 TO N1
42112 FOR I2=1 TO N2
42113 B(I1,I2)=C(I1,I2)
42114 NEXT I2
42115 NEXT I1
42116 RETURN
```

```
42117 IF N1=0 THEN RETURN
42118 FOR I1=1 TO N1
42119 B(I1)=C(I1)
42120 NEXT I1
42121 RETURN
42123 REM MATRIX SAVE (B IN C) SUBROUTINE (MATSAVBC)
42124 REM N1,N2 AND N3 ARE INPUT INDICES
42125 IF N1*N2*N3=0 THEN GOTO 42135
42126 REM CHECK DIMENSION
42127 FOR I1=1 TO N1
42128 FOR I2=1 TO N2
42129 FOR I3=1 TO N3
42130 C(I1,I2,I3)=B(I1,I2,I3)
42131 NEXT I3
42132 NEXT I2
42133 NEXT I1
42134 RETURN
42135 IF N1*N2=0 THEN GOTO 42142
42136 FOR I1=1 TO N1
42137 FOR I2=1 TO N2
42138 C(I1,I2)=B(I1,I2)
42139 NEXT I2
42140 NEXT I1
42141 RETURN
42142 IF N1=0 THEN RETURN
42143 FOR I1=1 TO N1
42144 C(I1)=B(I1)
42145 NEXT I1
42146 RETURN
42148 REM MATRIX SAVE (A IN C) SUBROUTINE (MATSAVAC)
42149 REM N1,N2 AND N3 ARE INPUT INDICES
42150 IF N1*N2*N3=0 THEN GOTO 42160
42151 REM CHECK DIMENSION
42152 FOR I1=1 TO N1
42153 FOR I2=1 TO N2
42154 FOR I3=1 TO N3
42155 C(I1,I2,I3)=A(I1,I2,I3)
42156 NEXT I3
42157 NEXT I2
42158 NEXT I1
42159 RETURN
42160 IF N1*N2=0 THEN GOTO 42167
42161 FOR I1=1 TO N1
42162 FOR I2=1 TO N2
42163 C(I1,I2)=A(I1,I2)
42164 NEXT I2
42165 NEXT I1
42166 RETURN
42167 IF N1=0 THEN RETURN
```

```
42168 FOR I1=1 TO N1
42169 C(I1)=A(I1)
42170 NEXT I1
42171 RETURN
42173 REM MATRIX SAVE (C IN A) SUBROUTINE (MATSAVCA)
42174 REM N1,N2 AND N3 ARE INPUT INDICES
42175 IF N1*N2*N3=0 THEN GOTO 42185
42176 REM CHECK DIMENSION
42177 FOR I1=1 TO N1
42178 FOR I2=1 TO N2
42179 FOR I3=1 TO N3
42180 A(I1,I2,I3)=C(I1,I2,I3)
42181 NEXT I3
42182 NEXT I2
42183 NEXT I1
42184 RETURN
42185 IF N1*N2=0 THEN GOTO 42192
42186 FOR I1=1 TO N1
42187 FOR I2=1 TO N2
42188 A(I1,I2)=C(I1,I2)
42189 NEXT I2
42190 NEXT I1
42191 RETURN
42192 IF N1=0 THEN RETURN
42193 FOR I1=1 TO N1
42194 A(I1)=C(I1)
42195 NEXT I1
42196 RETURN
42198 REM SCALAR B X MATRIX A SUBROUTINE (MATSCALE)
42199 REM N1,N2 AND N3 ARE INPUT INDICES
42200 IF N1*N2*N3=0 THEN GOTO 42210
42201 REM CHECK DIMENSION
42202 FOR I1=1 TO N1
42203 FOR I2=1 TO N2
42204 FOR I3=1 TO N3
42205 A(I1,I2,I3)=B*A(I1,I2,I3)
42206 NEXT I3
42207 NEXT I2
42208 NEXT I1
42209 RETURN
42210 IF N1*N2=0 THEN GOTO 42217
42211 FOR I1=1 TO N1
42212 FOR I2=1 TO N2
42213 A(I1,I2)=B*A(I1,I2)
42214 NEXT I2
42215 NEXT I1
42216 RETURN
42217 IF N1=0 THEN RETURN
42218 FOR I1=1 TO N1
```

```
42219 A(I1)=B*A(I1)
42220 NEXT I1
42221 RETURN
42223 REM MATRIX A CLEAR SUBROUTINE (MATCLRA)
42224 REM N1,N2 AND N3 ARE INPUT INDICES
42225 IF N1*N2*N3=0 THEN GOTO 42235
42226 REM CHECK DIMENSION
42227 FOR I1=1 TO N1
42228 FOR I2=1 TO N2
42229 FOR I3=1 TO N3
42230 A(I1,I2,I3)=0
42231 NEXT I3
42232 NEXT I2
42233 NEXT I1
42234 RETURN
42235 IF N1*N2=0 THEN GOTO 42242
42236 FOR I1=1 TO N1
42237 FOR I2=1 TO N2
42238 A(I1,I2)=0
42239 NEXT I2
42240 NEXT I1
42241 RETURN
42242 IF N1=0 THEN RETURN
42243 FOR I1=1 TO N1
42244 A(I1)=0
42245 NEXT I1
42246 RETURN
42248 REM ROW SWITCHING SUBROUTINE (MATSWCH)
42249 REM ROWS N1 AND N2 ARE INTERCHANGED
42250 FOR J=1 TO N
42251 B=A(N1,J)
42252 A(N1,J)=A(N2,J)
42253 A(N2,J)=B
42254 NEXT J
42255 RETURN
42273 REM ROW MULTIPLICATION/ADD SUBROUTINE (MATRMAD)
42274 REM B TIMES ROW N1 ADDED TO N2
42275 FOR J=1 TO N
42276 A(N2,J)=A(N2,J)+B*A(N1,J)
42277 NEXT J
42278 RETURN
42296 REM COFACTOR K SUBROUTINE (MATCOFAT)
42297 REM INPUT MATRIX SIZE IS N X N
42298 REM MATRIX A(I,J) IN, MATRIX B(I,J) OUT
42299 REM FIRST SHIFT UP ONE ROW
42300 FOR I=2 TO N
42301 FOR J=1 TO N
42302 B(I-1,J)=A(I,J)
42303 NEXT J
```

```
42304 NEXT I
42305 FOR I=1 TO N-1
42306 FOR J=K TO N
42307 IF K=N THEN GOTO 42309
42308 B(I,J)=B(I,J+1)
42309 NEXT J
42310 NEXT I
42311 RETURN
42348 REM MATRIX DETERMINANT SUBROUTINE (MATDET)
42349 REM FINDS DETERMINANT FOR UP TO A 4 X 4 MATRIX
42350 IF N>=2 THEN GOTO 42355
42351 REM ********************
42352 REM FIRST ORDER DETERMINANT
42353 D=A(1,1)
42354 RETURN
42355 IF N>=3 THEN GOTO 42360
42356 REM ********************
42357 REM SECOND ORDER DETERMINANT
42358 D=A(1,1)*A(2,2)-A(1,2)*A(2,1)
42359 RETURN
42360 IF N>=4 THEN GOTO 42370
42361 REM ********************
42362 REM THIRD ORDER DETERMINANT
42363 D=A(1,1)*(A(2,2)*A(3,3)-A(2,3)*A(3,2))
42364 D=D-A(1,2)*(A(2,1)*A(3,3)-A(2,3)*A(3,1))
42365 D=D+A(1,3)*(A(2,1)*A(3,2)-A(2,2)*A(3,1))
42366 RETURN
42367 REM ********************
42368 REM FOURTH ORDER DETERMINANT
42369 REM SAVE A IN C
42370 N1=N
42371 N2=N
42372 N3=0
42373 GOSUB 42150
42374 IF N>=5 THEN RETURN
42375 REM D1 WILL BE THE DETERMINANT
42376 D1=0
42377 REM FIND DETERMINANT OF EACH COFACTOR
42378 FOR K=1 TO 4
42379 REM GET COFACTOR K
42380 GOSUB 42300
42381 REM COFACTOR RETURNED IN B
42382 REM MOVE B TO A
42383 GOSUB 42075
42384 REM GET DET(A)
42385 GOSUB 42363
42386 D1=D1+C(1,K)*D
42387 REM REVERSE SIGN FOR NEXT COFACTOR
42388 D1=-D1
```

```
42389 REM SAVE C IN A
42390 GOSUB 42175
42391 NEXT K
42392 D=D1
42393 RETURN
42395 REM MATRIX INVERSION SUBROUTINE (MATINV)
42396 REM GAUSS-JORDAN ELIMINATION
42397 REM MATRIX A IS INPUT, MATRIX B IS OUTPUT
42398 REM DIM A=N X N    TEMPORARY DIM B=N X 2N
42399 REM FIRST CREATE MATRIX WITH A ON THE LEFT AND I ON THE RIGHT
42400 FOR I=1 TO N
42401 FOR J=1 TO N
42402 B(I,J+N)=0
42403 B(I,J)=A(I,J)
42404 NEXT J
42405 B(I,I+N)=1
42406 NEXT I
42407 REM PERFORM ROW ORIENTED OPERATIONS TO CONVERT THE LEFT HAND
42408 REM SIDE OF B TO THE IDENTITY MATRIX. THE INVERSE OF A WILL
42409 REM THEN BE ON THE RIGHT.
42410 FOR K=1 TO N
42411 IF K=N THEN GOTO 42424
42412 M=K
42413 REM FIND MAXIMUM ELEMENT
42414 FOR I=K+1 TO N
42415 IF ABS(B(I,K))>ABS(B(M,K)) THEN M=I
42416 NEXT I
42417 IF M=K THEN GOTO 42424
42418 FOR J=K TO 2*N
42419 B=B(K,J)
42420 B(K,J)=B(M,J)
42421 B(M,J)=B
42422 NEXT J
42423 REM DIVIDE ROW K
42424 FOR J=K+1 TO 2*N
42425 B(K,J)=B(K,J)/B(K,K)
42426 NEXT J
42427 IF K=1 THEN GOTO 42434
42428 FOR I=1 TO K-1
42429 FOR J=K+1 TO 2*N
42430 B(I,J)=B(I,J)-B(I,K)*B(K,J)
42431 NEXT J
42432 NEXT I
42433 IF K=N THEN GOTO 42441
42434 FOR I=K+1 TO N
42435 FOR J=K+1 TO 2*N
42436 B(I,J)=B(I,J)-B(I,K)*B(K,J)
42437 NEXT J
42438 NEXT I
```

```
42439 NEXT K
42440 REM RETRIEVE INVERSE FROM THE RIGHT SIDE OF B
42441 FOR I=1 TO N
42442 FOR J=1 TO N
42443 B(I,J)=B(I,J+N)
42444 NEXT J
42445 NEXT I
42446 RETURN
42693 REM EIGENVALUE (POWER METHOD) SUBROUTINE (EIGENPOW)
42694 REM AX=LX
42695 REM A IS THE N X N MATRIX
42696 REM B IS AN ARBITRARY VECTOR
42697 REM E IS THE RELATIVE ERROR CHOSEN
42698 REM  D= COUNT OF THE NUMBER OF ITERATIONS
42699 REM SET PARAMETERS NEEDED FOR MULTIPLY SUBROUTINE
42700 M1=N
42701 N1=N
42702 M2=N
42703 N2=1
42704 REM GENERATE ARBITRARY NORMALIZED VECTOR B(I,1)
42705 FOR I=1 TO N
42706 B(I,1)=1/SQRT(N)
42707 NEXT I
42708 REM B = LAST EIGENVALUE ESTIMATE
42709 REM A = CURRENT EIGENVALUE ESTIMATE
42710 REM PICK AN INITIAL VALUE FOR THE EIGENVALUE GUESS
42711 B=1
42712 D=0
42713 REM START ITERATION
42714 A=0
42715 GOSUB 41900
42716 REM CONVERT C OUTPUT TO B
42717 FOR I=1 TO N
42718 B(I,1)=C(I,1)
42719 A=A+B(I,1)*B(I,1)
42720 NEXT I
42721 D=D+1
42722 A=SQRT(A)
42723 REM NORMALIZE VECTOR
42724 FOR I=1 TO N
42725 B(I,1)=B(I,1)/A
42726 NEXT I
42727 IF ABS((A-B)/A)<E THEN RETURN
42728 B=A
42729 IF D>D1 THEN RETURN
42730 GOTO 42714
42796 REM MATRIX EXPONENT SUBROUTINE (MATEXP)
42797 REM INPUTS TO THE SUBROUTINE ARE THE MATRIX A, MATRIX
42798 REM SIZE N, NUMBER OF TERMS K2, AND VARIABLE X
```

```
42799 REM SET UP INDICES TO BE USED LATER
42800 N1=N
42801 N2=N
42802 N3=0
42803 M1=N
42804 M2=N
42805 REM GUARD AGAINST DIVIDE BY ZERO
42806 IF X=0 THEN X=.0000000000001
42807 REM INITIALIZE STORAGE MATRIX D(I,J)
42808 FOR I=1 TO N
42809 FOR J=1 TO N
42810 D(I,J)=0
42811 NEXT J
42812 NEXT I
42813 REM K2 IS THE NUMBER OF TERMS TO BE CALCULATED
42814 K1=0
42815 REM CREATE IDENTITY MATRIX IN B
42816 B=1
42817 GOSUB 42000
42818 REM MOVE B TO C
42819 GOSUB 42125
42820 REM ADD TO D
42821 GOSUB 42847
42822 K1=K1+1
42823 IF K1>=K2 THEN GOTO 42838
42824 REM SCALE MATRIX A BY X/K1
42825 B=X/K1
42826 GOSUB 42200
42827 REM MULTIPLY A TIMES B
42828 GOSUB 41900
42829 REM ADD RESULT TO MATRIX D
42830 GOSUB 42847
42831 REM MOVE C TO B
42832 GOSUB 42100
42833 REM RETURN MATRIX A TO ORIGINAL CONDITION
42834 B=K1/X
42835 GOSUB 42200
42836 REM CONTINUE SUMMATION
42837 GOTO 42822
42838 REM MOVE RESULT IN D TO C
42839 FOR I=1 TO N
42840 FOR J=1 TO N
42841 C(I,J)=D(I,J)
42842 NEXT J
42843 NEXT I
42844 REM RETURN TO CALLING PROGRAM
42845 RETURN
42846 REM D(I,J) IS USED FOR TEMPORARY STORAGE
42847 FOR I=1 TO N
```

```
42848 FOR J=1 TO N
42849 D(I,J)=D(I,J)+C(I,J)
42850 NEXT J
42851 NEXT I
42852 RETURN
42899 REM LINEAR RANDOM NUMBER GENERATOR (LINEAR)
42900 REM U=MEAN, V=SPREAD, D=SEED
42901 I9=I9+1
42902 A=3.14159265358979323846
42903 B=2.71828182845904523536
42904 C=1.41421356237309504880
42905 D=1+ABS(D)
42906 E=E+(1+D/B)*C
42907 E=E*I9
42908 E=E-A*INT(E/A)
42909 E=E-INT(E)+.018
42910 IF E>.1 THEN E=E+.009
42911 IF E>.2 THEN E=E-.002
42912 IF E>.3 THEN E=E-.005
42913 IF E>.4 THEN E=E-.005
42914 IF E>.5 THEN E=E-.015
42915 E=V*(E-.5)+U
42916 RETURN
42923 REM NORMAL DISTRIBUTION BY CENTRAL LIMIT THEOREM (NORMAL)
42924 REM U=MEAN, V=STANDARD DEVIATION, E=RANDOM NO. GENERATED
42925 E=0
42926 FOR I9=1 TO 48
42927 E=E+RND(0)-.5
42928 NEXT I9
42929 E=V*E/2+U
42930 RETURN
42948 REM POISSON RANDOM NUMBER GENERATOR (POISSON)
42949 REM INPUT PARAMETER U
42950 X=RND(0)*EXP(U)
42951 X1=1
42952 Y1=1
42953 Y=0
42954 IF X1>X THEN GOTO 42959
42955 Y=Y+1
42956 Y1=Y1*U/Y
42957 X1=X1+Y1
42958 GOTO 42954
42959 IF Y>0 THEN Y=Y-(X1-X)/Y1
42960 E=Y
42961 RETURN
42972 REM BINOMIAL RANDOM NUMBER GENERATOR (BINOMIAL)
42973 REM B=PROBABILITY PER TRIAL
42974 REM N=NUMBER OF TRIALS
42975 E=0
```

```
42976 FOR K=1 TO N
42977 Y1=RND(0)
42978 IF Y1<B THEN E=E+1
42979 NEXT K
42980 RETURN
42998 REM EXPONENTIAL RANDOM NUMBER GENERATOR (EXPONENT)
42999 REM U=MEAN
43000 X=RND(0)
43001 E=-U*LOG(1-X)
43002 RETURN
43023 REM FERMI RANDOM NUMBER GENERATOR (FERMI)
43024 REM U=INFLECTION POINT, V=SPREAD OF TRANSITION REGION
43025 X=RND(0)
43026 Y=1
43027 A=EXP(4*U/V)
43028 B=(X-1)*LOG(1+A)
43029 Y1=B+LOG(A+EXP(Y))
43030 IF ABS((Y-Y1)/Y)<.001 THEN GOTO 43033
43031 Y=Y1
43032 GOTO 43029
43033 E=V*Y1/4
43034 RETURN
43048 REM CAUCHY RANDOM NUMBER GENERATOR (CAUCHY)
43049 REM U=MEAN
43050 X=RND(0)
43051 E=U*SIN(1.5707963267*X)/COS(1.5707963267*X)
43052 RETURN
43073 REM GAMMA (N=2) RANDOM NUMBER GENERATOR (GAMMA)
43074 REM B=INPUT PARAMETER
43075 Y=1
43076 X=RND(0)
43077 Y1=-LOG((1-X)/(1+Y))
43078 IF ABS((Y1-Y)/Y)<.001 THEN GOTO 43081
43079 Y=Y1
43080 GOTO 43077
43081 E=B*Y1
43082 RETURN
43096 REM BETA RANDOM NUMBER GENERATOR (BETA)
43097 REM INPUT PARAMETERS ARE A AND B
43098 REM A IS RESTRICTED TO A=1 AND A=2
43099 REM GUARD AGAINST DIVIDE BY ZERO
43100 IF B>0 THEN GOTO 43103
43101 E=1
43102 RETURN
43103 REM B>0
43104 IF A>2 THEN RETURN
43105 X=RND(0)
43106 IF A=2 THEN GOTO 43109
43107 E=1-(1-X)^(1/B)
```

```
43108 RETURN
43109 Y=1
43110 Y1=1-((1-X)/(1+B*Y))^(1/B)
43111 IF ABS((Y-Y1)/Y)<.001 THEN GOTO 43114
43112 Y=Y1
43113 GOTO 43110
43114 E=Y1
43115 RETURN
43148 REM WEIBULL RANDOM NUMBER GENERATOR (WEIBULL)
43149 REM INPUT PARAMETERS ARE U AND V
43150 X=RND(0)
43151 E=U*((LOG(1/(1-X)))^(1/V))
43152 RETURN
43199 REM SERIES SUMMATION SUBROUTINE (SERSUM)
43200 Y=0
43201 FOR I=N TO 0 STEP -1
43202 Y=Z*Y+A(I)
43203 NEXT I
43204 RETURN
43209 REM SINE SERIES SUBROUTINE (SINE)
43210 X1=1
43211 IF X<0 THEN X1=-1
43212 X3=ABS(X)
43213 X2=3.141592653589793
43214 REM REDUCE RANGE
43215 X3=X3-2*X2*INT(.5*X3/X2)
43216 IF X3>X2 THEN X1=-X1
43217 IF X3>X2 THEN X3=X3-X2
43218 IF X3>X2/2 THEN X3=X2-X3
43219 Z=X3*X3
43220 GOSUB 43300
43221 GOSUB 43200
43222 Y=X1*X3*Y
43223 RETURN
43224 REM COSINE SERIES SUBROUTINE (COSINE)
43225 X2=3.141592653589793
43226 X1=1
43227 REM REDUCE RANGE
43228 X3=ABS(X)
43229 X3=X3-2*X2*INT(0.5*X3/X2)
43230 IF X3>X2 THEN X1=-1
43231 IF X3>X2 THEN X3=X3-X2
43232 IF X3>X2/2 THEN X1=-X1
43233 IF X3>X2/2 THEN X3=X2-X3
43234 Z=X3*X3
43235 GOSUB 43310
43236 GOSUB 43200
43237 Y=X1*Y
43238 RETURN
```

```
43244 REM ARCTANGENT SERIES SUBROUTINE (ARCTAN)
43245 X1=1
43246 IF X<0 THEN X1=-1
43247 X3=ABS(X)
43248 Z1=(X3-1)/(X3+1)
43249 IF X3<.5 THEN Z1=X3
43250 Z=Z1*Z1
43251 REM GET SERIES COEFFICIENTS
43252 GOSUB 43320
43253 REM SUM SERIES
43254 GOSUB 43200
43255 Y=Z1*Y
43256 IF X3>=.5 THEN Y=Y+.78539816339745
43257 Y=X1*Y
43258 RETURN
43259 REM LOG BASE TEN SERIES SUBROUTINE (LOG(X))
43260 X1=1
43261 C=-1
43262 X2=10
43263 X3=X
43264 IF X>=1 THEN GOTO 43268
43265 X=1/X
43266 X1=-1
43267 REM REDUCE RANGE
43268 X=X2*X
43269 C=C+1
43270 X=X/X2
43271 IF X>X2 THEN GOTO 43269
43272 Z1=(X-3.162277660)/(X+3.162277660)
43273 Z=Z1*Z1
43274 GOSUB 43340
43275 GOSUB 43200
43276 Y=X1*(C+Z1*Y+.5)
43277 X=X3
43278 RETURN
43279 REM NATURAL LOGARITHM SERIES SUBROUTINE (LN(X))
43280 X1=1
43281 C=-1
43282 X2=2.718281828459045
43283 X3=X
43284 IF X>=1 THEN GOTO 43288
43285 X=1/X
43286 X1=-1
43287 REM REDUCE RANGE
43288 X=X2*X
43289 C=C+1
43290 X=X/X2
43291 IF X>X2 THEN GOTO 43289
43292 Z1=(X-1.6487212707)/(X+1.6487212707)
```

```
43293 Z=Z1*Z1
43294 GOSUB 43360
43295 GOSUB 43200
43296 Y=X1*(C+Z1*Y+.5)
43297 X=X3
43298 RETURN
43299 REM SINE SERIES COEFFICIENTS (SINDATA)
43300 N=6
43301 A(0)=1
43302 A(1)=-.1666666666671334
43303 A(2)=.00833333333809067
43304 A(3)=-.0001984127155551283
43305 A(4)=.0000027557589750762
43306 A(5)=-.00000002507059876207
43307 A(6)=.000000000164105986683
43308 RETURN
43309 REM COSINE SERIES COEFFICIENTS (COSDATA)
43310 N=6
43311 A(0)=1
43312 A(1)=-.49999999999982
43313 A(2)=.04166666664651
43314 A(3)=-.001388888805755
43315 A(4)=.000024801428034
43316 A(5)=-.0000002754213324
43317 A(6)=.0000000020189405
43318 RETURN
43319 REM ARCTANGENT SERIES COEFFICIENTS (ARCDATA)
43320 N=10
43321 A(0)=1
43322 A(1)=-.333333311286
43323 A(2)=.199998774421
43324 A(3)=-.142831560376
43325 A(4)=.110840091104
43326 A(5)=-.089229124381
43327 A(6)=.070315200033
43328 A(7)=-.049278908030
43329 A(8)=.026879941561
43330 A(9)=-.009568384520
43331 A(10)=.001605444922
43332 RETURN
43339 REM LOG BASE TEN SERIES COEFFICIENTS (LOGDATA)
43340 N=9
43341 A(0)=.8685889644
43342 A(1)=.2895297117
43343 A(2)=.1737120608
43344 A(3)=.1242584742
43345 A(4)=.0939080460
43346 A(5)=.1009301264
43347 A(6)=-.0439630355
```

```
43348 A(7)=.3920576195
43349 A(8)=-.5170494708
43350 A(9)=.4915571108
43351 RETURN
43359 REM NATURAL LOGARITHM SERIES COEFFICIENTS (LNDATA)
43360 N=9
43361 A(0)=2
43362 A(1)=.66666672443
43363 A(2)=.3999895288
43364 A(3)=.286436047
43365 A(4)=.197959107
43366 A(5)=.6283533
43367 A(6)=-4.54692
43368 A(7)=28.117
43369 A(8)=-86.42
43370 A(9)=106.1
43371 RETURN
43379 REM POWER OF E SERIES COEFFICIENTS (EXPDATA)
43380 N=8
43381 A(0)=1
43382 A(1)=.99999999668
43383 A(2)=.49999995173
43384 A(3)=.16666704243
43385 A(4)=.04166685027
43386 A(5)=.00832672635
43387 A(6)=.00140836136
43388 A(7)=.00017358267
43389 A(8)=.00003931683
43390 RETURN
43399 REM POWER OF TEN SERIES COEFFICIENTS (TENDATA)
43400 N=9
43401 A(0)=1
43402 A(1)=1.1512925485
43403 A(2)=.6627373050
43404 A(3)=.2543345675
43405 A(4)=.0732032741
43406 A(5)=.0168603036
43407 A(6)=.0032196227
43408 A(7)=.0005547662
43409 A(8)=.0000573305
43410 A(9)=.0000179419
43411 RETURN
43449 REM POWER OF TEN SERIES SUBROUTINE (TENPOW)
43450 X1=1
43451 X3=X
43452 IF X<0 THEN X1=-1
43453 X=ABS(X)
43454 REM REDUCE RANGE
43455 X2=INT(X)
```

```
43456 X=X-X2
43457 REM GET COEFFICIENTS
43458 GOSUB 43400
43459 Z=X
43460 REM SUM SERIES
43461 GOSUB 43200
43462 Y=Y*Y
43463 FOR I=1 TO X2
43464 Y=Y*10
43465 NEXT I
43466 IF X1<0 THEN Y=1/Y
43467 X=X3
43468 RETURN
43469 REM EXPONENT SERIES SUBROUTINE (EXP(X))
43470 X1=1
43471 X3=X
43472 IF X<0 THEN X1=-1
43473 X=ABS(X)
43474 REM REDUCE RANGE
43475 X2=INT(X)
43476 X=X-X2
43477 REM GET COEFFICIENTS
43478 GOSUB 43380
43479 Z=X
43480 REM SUM SERIES
43481 GOSUB 43200
43482 FOR I=1 TO X2
43483 Y=Y*2.718281828459045
43484 NEXT I
43485 IF X1<0 THEN Y=1/Y
43486 X=X3
43487 RETURN
```

APPENDIX IIB

Compacted Subroutine Listing

APPENDIX IIB *Compacted Subroutine Listing*

The subroutine library shown in Appendix IIA was compacted by removing the unnecessary REMark statements and spaces.

Compacted subroutine library length: 10759 bytes

```
40000B=100000000
40002FORI=1TON
40003IFB>D(I)THENB=D(I)
40004NEXTI
40006FORI=1TON
40007D(I)=D(I)-B
40008NEXTI
40010C=0
40011FORI=1TON
40012IFC<D(I)THENC=D(I)
40013NEXTI
40015A=L/C
40017E=A*ABS(B)
40018PRINT
40019PRINT
40020PRINT"***** DATA PLOT (SCALED) *****"
40021PRINT
40022PRINT
40023PRINT"MIN. ORDINATE=  ",B,"   MAX. ORDINATE= ",C+B
40024PRINT"INITIAL ABSCISSA VALUE= ",X1
40025PRINT
40026PRINT
40028IFB>0THENGOTO40034
40030IFABS(B)>CTHENGOTO40034
40032PRINTTAB(E),"0"
40034GOSUB40200
40035FORI=1TON
40037FORK=1TO(INT(0.6*L/N))
40038PRINT":",TAB(L),":"
40039NEXTK
40041E2=A*D(I)
40043IFE2>=1THENGOTO40046
```

```
40044PRINT"*",
40045GOTO40049
40046PRINT":",
40047PRINTTAB(E2),"*",
40048IFINT(E2)=LTHENGOTO40050
40049PRINTTAB(L),":",
40050PRINT
40051NEXTI
40053GOSUB40200
40054PRINT
40055PRINT
40056PRINT"END ABSCISSA VALUE= ",X2
40057PRINT
40058PRINT
40060RETURN
40100B=100000000
40102FORI=1TON
40103IFB>D(I)THENB=D(I)
40104NEXTI
40106FORI=1TON
40107D(I)=D(I)-B
40108NEXTI
40110C=0
40111FORI=1TON
40112IFC<D(I)THENC=D(I)
40113NEXTI
40115E(0)=0
40116E(N)=1
40117FORI=2TON
40118E(I-1)=INT(0.5*(C(I)-C(I-1))*L/(C(N)-C(1))+.5)
40120NEXTI
40122A=L/C
40124E=A*ABS(B)
40125PRINT
40126PRINT
40127PRINT"***** DATA PLOT (SCALED) *****"
40128PRINT
40129PRINT
40130PRINT"MIN. ORDINATE=  ",B,"    MAX. ORDINATE= ",C+B
40131PRINT"INITIAL ABSCISSA VALUE= ",C(1)
40132PRINT
40133PRINT
40135IFB>0THENGOTO40141
40137IFABS(B)>CTHENGOTO40141
40139PRINTTAB(E),"0"
40141GOSUB40200
40142FORI=1TON
40144FORK=1TOE(I-1)
40145PRINT":",TAB(L),":"
```

```
40146NEXTK
40148E2=A*D(I)
40150IFE(I)=0THENGOTO40161
40152IFE2>=1THENGOTO40155
40153PRINT"*",
40154GOTO40158
40155PRINT":",
40156PRINTTAB(E2),"*",
40157IFINT(E2)=LTHENGOTO40159
40158PRINTTAB(L),":",
40159PRINT
40160GOTO40175
40161 TEST FOR OVERLAY OF POINTS
40163IFD(I)=D(I+1)THENGOTO40175
40164IFE2>=1THENGOTO40168
40166PRINT"*",TAB(A*D(I+1)),"*",
40167GOTO40171
40168PRINT":",
40169PRINTTAB(E2),"*",TAB(A*D(I+1)),"*",
40171IFINT(A*D(I+1))=LTHENGOTO40173
40172PRINTTAB(L),":",
40173PRINT
40174I=I+1
40175NEXTI
40177GOSUB40200
40178PRINT
40179PRINT
40180PRINT"END ABSCISSA VALUE= ",C(N)
40181PRINT
40182PRINT
40184RETURN
40200E3=E-5*INT(E/5)
40202IFB>0THENE3=0
40204IFABS(B)>CTHENE3=0
40205FORJ=1TOE3
40206PRINT"-",
40207NEXTJ
40208FORJ=1TO(L-E3)/5
40209PRINT"I----",
40210NEXTJ
40211PRINT"I",
40212E4=(J-1)*5+1+E3
40213IFE4=L+1THENPRINT
40214IFE4=L+1THENGOTO40221
40215E4=E4+1
40216IFE4>=L+1THENGOTO40219
40217PRINT"-",
40218GOTO40215
40219PRINT":"
```

```
40221RETURN
40300X(3)=X(1)+X(2)
40301Y(3)=Y(1)+Y(2)
40302RETURN
40350X(3)=X(1)-X(2)
40351Y(3)=Y(1)-Y(2)
40352RETURN
40400U=SQRT(X*X+Y*Y)
40402IFY=0THENY=(.1)^30
40404IFX=0THENX=(.1)^30
40406W=Y/X
40407V=ATN(W)
40409IFX<0THENV=V+3.1415926535
40410IFV<0THENV=V+6.2831853072
40411RETURN
40450X=U*COS(V)
40451Y=U*SIN(V)
40452RETURN
40500U=U(1)*U(2)
40501V=V(1)+V(2)
40502IFV>=6.2831853072THENV=V-6.2831853072
40503RETURN
40550U=U(1)/U(2)
40551V=V(1)-V(2)
40552IFV<0THENV=V+6.2831853072
40553RETURN
40600X=X(1)
40601Y=Y(1)
40603GOSUB40400
40604U(1)=U
40605V(1)=V
40606X=X(2)
40607Y=Y(2)
40609GOSUB40400
40610U(2)=U
40611V(2)=V
40613GOSUB40500
40615GOSUB40450
40616RETURN
40800X=X(1)
40801Y=Y(1)
40803GOSUB40400
40804U(1)=U
40805V(1)=V
40806X=X(2)
40807Y=Y(2)
40809GOSUB40400
40810U(2)=U
40811V(2)=V
```

```
40813GOSUB40550
40815GOSUB40450
40816RETURN
41100U1=U^N
41101V1=N*V
41102V1=V1-6.2831853072*INT(V1/6.2831853072)
41103RETURN
41150U1=U^(1/N)
41151V1=V/N
41152RETURN
41200GOSUB40400
41202GOSUB41100
41204U=U1
41205V=V1
41207GOSUB40450
41208RETURN
41300GOSUB40400
41302GOSUB41150
41303U=U1
41306V=V1+6.2831853072*(M-1)/N
41308GOSUB40450
41309RETURN
41400X=U*(SIN(W))*COS(V)
41401Y=U*(SIN(W))*SIN(V)
41402Z=U*COS(W)
41403RETURN
41450GOSUB40400
41452V1=V
41453X=U
41454Y=Z
41456GOSUB40400
41457IFV>1.5707963268THENV=V-6.28318553072
41458W=1.5707963268-V
41459V=V1
41460RETURN
41500FORI=1TON
41501C(I)=A(I)+B(I)
41502NEXTI
41503RETURN
41550FORI=1TON
41551C(I)=A(I)-B(I)
41552NEXTI
41553RETURN
41600C=0
41601FORI=1TON
41602C=C+A(I)*B(I)
41603NEXTI
41604RETURN
41650C(1)=A(2)*B(3)-A(3)*B(2)
```

```
41651C(2)=A(3)*B(1)-A(1)*B(3)
41652C(3)=A(1)*B(2)-A(2)*B(1)
41653RETURN
41700L=0
41701FORI=1TON
41702L=L+A(I)*A(I)
41703NEXTI
41704L=SQRT(L)
41705RETURN
41750GOSUB41600
41752GOSUB41700
41754L1=L
41756FORI=1TON
41757A(I)=B(I)
41758NEXTI
41759GOSUB41700
41760E=C/(L*L1)+(.1)^30
41761E=SQRT(1-E*E)/E
41762A=ATN(E)
41763IFC<0THENA=3.1415926536-A
41764RETURN
41800FORI=1TOM
41801FORJ=1TON
41802C(I,J)=A(I,J)+B(I,J)
41803NEXTJ
41804NEXTI
41805RETURN
41850FORI=1TOM
41851FORJ=1TON
41852C(I,J)=A(I,J)-B(I,J)
41853NEXTJ
41854NEXTI
41855RETURN
41900FORI=1TOM1
41901FORJ=1TON2
41902C(I,J)=0
41903FORK=1TON1
41904C(I,J)=C(I,J)+A(I,K)*B(K,J)
41905NEXTK
41906NEXTJ
41907NEXTI
41908RETURN
41950FORI=1TON
41951FORJ=1TOM
41952B(I,J)=A(J,I)
41953NEXTJ
41954NEXTI
41955RETURN
42000FORI=1TON
```

```
42001FORJ=1TON
42002B(I,J)=0
42003IFI=JTHENB(I,J)=B
42004NEXTJ
42005NEXTI
42006RETURN
42050IFN1*N2*N3=0THENGOTO42060
42052FORI1=1TON1
42053FORI2=1TON2
42054FORI3=1TON3
42055B(I1,I2,I3)=A(I1,I2,I3)
42056NEXTI3
42057NEXTI2
42058NEXTI1
42059RETURN
42060IFN1*N2=0THENGOTO42067
42061FORI1=1TON1
42062FORI2=1TON2
42063B(I1,I2)=A(I1,I2)
42064NEXTI2
42065NEXTI1
42066RETURN
42067IFN1=0THENRETURN
42068FORI1=1TON1
42069B(I1)=A(I1)
42070NEXTI1
42071RETURN
42075IFN1*N2*N3=0THENGOTO42085
42077FORI1=1TON1
42078FORI2=1TON2
42079FORI3=1TON3
42080A(I1,I2,I3)=B(I1,I2,I3)
42081NEXTI3
42082NEXTI2
42083NEXTI1
42084RETURN
42085IFN1*N2=0THENGOTO42092
42086FORI1=1TON1
42087FORI2=1TON2
42088A(I1,I2)=B(I1,I2)
42089NEXTI2
42090NEXTI1
42091RETURN
42092IFN1=0THENRETURN
42093FORI1=1TON1
42094A(I1)=B(I1)
42095NEXTI1
42096RETURN
42100IFN1*N2*N3=0THENGOTO42110
```

```
42102 FOR I1=1 TO N1
42103 FOR I2=1 TO N2
42104 FOR I3=1 TO N3
42105 B(I1,I2,I3)=C(I1,I2,I3)
42106 NEXT I3
42107 NEXT I2
42108 NEXT I1
42109 RETURN
42110 IF N1*N2=0 THEN GOTO 42117
42111 FOR I1=1 TO N1
42112 FOR I2=1 TO N2
42113 B(I1,I2)=C(I1,I2)
42114 NEXT I2
42115 NEXT I1
42116 RETURN
42117 IF N1=0 THEN RETURN
42118 FOR I1=1 TO N1
42119 B(I1)=C(I1)
42120 NEXT I1
42121 RETURN
42125 IF N1*N2*N3=0 THEN GOTO 42135
42127 FOR I1=1 TO N1
42128 FOR I2=1 TO N2
42129 FOR I3=1 TO N3
42130 C(I1,I2,I3)=B(I1,I2,I3)
42131 NEXT I3
42132 NEXT I2
42133 NEXT I1
42134 RETURN
42135 IF N1*N2=0 THEN GOTO 42142
42136 FOR I1=1 TO N1
42137 FOR I2=1 TO N2
42138 C(I1,I2)=B(I1,I2)
42139 NEXT I2
42140 NEXT I1
42141 RETURN
42142 IF N1=0 THEN RETURN
42143 FOR I1=1 TO N1
42144 C(I1)=B(I1)
42145 NEXT I1
42146 RETURN
42150 IF N1*N2*N3=0 THEN GOTO 42160
42152 FOR I1=1 TO N1
42153 FOR I2=1 TO N2
42154 FOR I3=1 TO N3
42155 C(I1,I2,I3)=A(I1,I2,I3)
42156 NEXT I3
42157 NEXT I2
42158 NEXT I1
```

```
42159RETURN
42160IFN1*N2=0THENGOTO42167
42161FORI1=1TON1
42162FORI2=1TON2
42163C(I1,I2)=A(I1,I2)
42164NEXTI2
42165NEXTI1
42166RETURN
42167IFN1=0THENRETURN
42168FORI1=1TON1
42169C(I1)=A(I1)
42170NEXTI1
42171RETURN
42175IFN1*N2*N3=0THENGOTO42185
42177FORI1=1TON1
42178FORI2=1TON2
42179FORI3=1TON3
42180A(I1,I2,I3)=C(I1,I2,I3)
42181NEXTI3
42182NEXTI2
42183NEXTI1
42184RETURN
42185IFN1*N2=0THENGOTO42192
42186FORI1=1TON1
42187FORI2=1TON2
42188A(I1,I2)=C(I1,I2)
42189NEXTI2
42190NEXTI1
42191RETURN
42192IFN1=0THENRETURN
42193FORI1=1TON1
42194A(I1)=C(I1)
42195NEXTI1
42196RETURN
42200IFN1*N2*N3=0THENGOTO42210
42202FORI1=1TON1
42203FORI2=1TON2
42204FORI3=1TON3
42205A(I1,I2,I3)=B*A(I1,I2,I3)
42206NEXTI3
42207NEXTI2
42208NEXTI1
42209RETURN
42210IFN1*N2=0THENGOTO42217
42211FORI1=1TON1
42212FORI2=1TON2
42213A(I1,I2)=B*A(I1,I2)
42214NEXTI2
42215NEXTI1
```

```
42216RETURN
42217IFN1=0THENRETURN
42218FORI1=1TON1
42219A(I1)=B*A(I1)
42220NEXTI1
42221RETURN
42225IFN1*N2*N3=0THENGOTO42235
42227FORI1=1TON1
42228FORI2=1TON2
42229FORI3=1TON3
42230A(I1,I2,I3)=0
42231NEXTI3
42232NEXTI2
42233NEXTI1
42234RETURN
42235IFN1*N2=0THENGOTO42242
42236FORI1=1TON1
42237FORI2=1TON2
42238A(I1,I2)=0
42239NEXTI2
42240NEXTI1
42241RETURN
42242IFN1=0THENRETURN
42243FORI1=1TON1
42244A(I1)=0
42245NEXTI1
42246RETURN
42250FORJ=1TON
42251B=A(N1,J)
42252A(N1,J)=A(N2,J)
42253A(N2,J)=B
42254NEXTJ
42255RETURN
42275FORJ=1TON
42276A(N2,J)=A(N2,J)+B*A(N1,J)
42277NEXTJ
42278RETURN
42300FORI=2TON
42301FORJ=1TON
42302B(I-1,J)=A(I,J)
42303NEXTJ
42304NEXTI
42305FORI=1TON-1
42306FORJ=KTON
42307IFK=NTHENGOTO42309
42308B(I,J)=B(I,J+1)
42309NEXTJ
42310NEXTI
42311RETURN
```

```
42350IFN>=2THENGOTO42355
42353D=A(1,1)
42354RETURN
42355IFN>=3THENGOTO42360
42358D=A(1,1)*A(2,2)-A(1,2)*A(2,1)
42359RETURN
42360IFN>=4THENGOTO42370
42363D=A(1,1)*(A(2,2)*A(3,3)-A(2,3)*A(3,2))
42364D=D-A(1,2)*(A(2,1)*A(3,3)-A(2,3)*A(3,1))
42365D=D+A(1,3)*(A(2,1)*A(3,2)-A(2,2)*A(3,1))
42366RETURN
42370N1=N
42371N2=N
42372N3=0
42373GOSUB42150
42374IFN>=5THENRETURN
42376D1=0
42378FORK=1TO4
42380GOSUB42300
42383GOSUB42075
42385GOSUB42363
42386D1=D1+C(1,K)*D
42388D1=-D1
42390GOSUB42175
42391NEXTK
42392D=D1
42393RETURN
42400FORI=1TON
42401FORJ=1TON
42402B(I,J+N)=0
42403B(I,J)=A(I,J)
42404NEXTJ
42405B(I,I+N)=1
42406NEXTI
42410FORK=1TON
42411IFK=NTHENGOTO42424
42412M=K
42414FORI=K+1TON
42415IFABS(B(I,K))>ABS(B(M,K))THENM=I
42416NEXTI
42417IFM=KTHENGOTO42424
42418FORJ=KTO2*N
42419B=B(K,J)
42420B(K,J)=B(M,J)
42421B(M,J)=B
42422NEXTJ
42424FORJ=K+1TO2*N
42425B(K,J)=B(K,J)/B(K,K)
42426NEXTJ
```

```
42427IFK=1THENGOTO42434
42428FORI=1TOK-1
42429FORJ=K+1TO2*N
42430B(I,J)=B(I,J)-B(I,K)*B(K,J)
42431NEXTJ
42432NEXTI
42433IFK=NTHENGOTO42441
42434FORI=K+1TON
42435FORJ=K+1TO2*N
42436B(I,J)=B(I,J)-B(I,K)*B(K,J)
42437NEXTJ
42438NEXTI
42439NEXTK
42441FORI=1TON
42442FORJ=1TON
42443B(I,J)=B(I,J+N)
42444NEXTJ
42445NEXTI
42446RETURN
42700M1=N
42701N1=N
42702M2=N
42703N2=1
42705FORI=1TON
42706B(I,1)=1/SQRT(N)
42707NEXTI
42711B=1
42712D=0
42714A=0
42715GOSUB41900
42717FORI=1TON
42718B(I,1)=C(I,1)
42719A=A+B(I,1)*B(I,1)
42720NEXTI
42721D=D+1
42722A=SQRT(A)
42724FORI=1TON
42725B(I,1)=B(I,1)/A
42726NEXTI
42727IFABS((A-B)/A)<ETHENRETURN
42728B=A
42729IFD>D1THENRETURN
42730GOTO42714
42800N1=N
42801N2=N
42802N3=0
42803M1=N
42804M2=N
42806IFX=0THENX=.000000000001
```

```
42808FORI=1TON
42809FORJ=1TON
42810D(I,J)=0
42811NEXTJ
42812NEXTI
42814K1=0
42816B=1
42817GOSUB42000
42819GOSUB42125
42821GOSUB42847
42822K1=K1+1
42823IFK1>=K2THENGOTO42838
42825B=X/K1
42826GOSUB42200
42828GOSUB41900
42830GOSUB42847
42832GOSUB42100
42834B=K1/X
42835GOSUB42200
42837GOTO42822
42839FORI=1TON
42840FORJ=1TON
42841C(I,J)=D(I,J)
42842NEXTJ
42843NEXTI
42845RETURN
42847FORI=1TON
42848FORJ=1TON
42849D(I,J)=D(I,J)+C(I,J)
42850NEXTJ
42851NEXTI
42852RETURN
42900 REM U=MEAN, V=SPREAD, D=SEED
42901I9=I9+1
42902A=3.14159265358979323846
42903B=2.71828182845904523536
42904C=1.41421356237309504880
42905D=1+ABS(D)
42906E=E+(1+D/B)*C
42907E=E*I9
42908E=E-A*INT(E/A)
42909E=E-INT(E)+.018
42910IFE>.1THENE=E+.009
42911IFE>.2THENE=E-.002
42912IFE>.3THENE=E-.005
42913IFE>.4THENE=E-.005
42914IFE>.5THENE=E-.015
42915E=V*(E-.5)+U
42916RETURN
```

```
42925E=0
42926FORI9=1TO48
42927E=E+RND(0)-.5
42928NEXTI9
42929E=V*E/2+U
42930RETURN
42950X=RND(0)*EXP(U)
42951X1=1
42952Y1=1
42953Y=0
42954IFX1>XTHENGOTO42959
42955Y=Y+1
42956Y1=Y1*U/Y
42957X1=X1+Y1
42958GOTO42954
42959IFY>0THENY=Y-(X1-X)/Y1
42960E=Y
42961RETURN
42975E=0
42976FORK=1TON
42977Y1=RND(0)
42978IFY1<BTHENE=E+1
42979NEXTK
42980RETURN
43000X=RND(0)
43001E=-U*LOG(1-X)
43002RETURN
43025X=RND(0)
43026Y=1
43027A=EXP(4*U/V)
43028B=(X-1)*LOG(1+A)
43029Y1=B+LOG(A+EXP(Y))
43030IFABS((Y-Y1)/Y)<.001THENGOTO43033
43031Y=Y1
43032GOTO43029
43033E=V*Y1/4
43034RETURN
43050X=RND(0)
43051E=U*SIN(1.5707963267*X)/COS(1.5707963267*X)
43052RETURN
43075Y=1
43076X=RND(0)
43077Y1=-LOG((1-X)/(1+Y))
43078IFABS((Y1-Y)/Y)<.001THENGOTO43081
43079Y=Y1
43080GOTO43077
43081E=B*Y1
43082RETURN
43100IFB>0THENGOTO43103
```

```
43101E=1
43102RETURN
43104IFA>2THENRETURN
43105X=RND(0)
43106IFA=2THENGOTO43109
43107E=1-(1-X)^(1/B)
43108RETURN
43109Y=1
43110Y1=1-((1-X)/(1+B*Y))^(1/B)
43111IFABS((Y-Y1)/Y)<.001THENGOTO43114
43112Y=Y1
43113GOTO43110
43114E=Y1
43115RETURN
43150X=RND(0)
43151E=U*((LOG(1/(1-X)))^(1/V))
43152RETURN
43200Y=0
43201FORI=NTO0STEP-1
43202Y=Z*Y+A(I)
43203NEXTI
43204RETURN
43210X1=1
43211IFX<0THENX1=-1
43212X3=ABS(X)
43213X2=3.141592653589793
43215X3=X3-2*X2*INT(.5*X3/X2)
43216IFX3>X2THENX1=-X1
43217IFX3>X2THENX3=X3-X2
43218IFX3>X2/2THENX3=X2-X3
43219Z=X3*X3
43220GOSUB43300
43221GOSUB43200
43222Y=X1*X3*Y
43223RETURN
43225X2=3.141592653589793
43226X1=1
43228X3=ABS(X)
43229X3=X3-2*X2*INT(0.5*X3/X2)
43230IFX3>X2THENX1=-1
43231IFX3>X2THENX3=X3-X2
43232IFX3>X2/2THENX1=-X1
43233IFX3>X2/2THENX3=X2-X3
43234Z=X3*X3
43235GOSUB43310
43236GOSUB43200
43237Y=X1*Y
43238RETURN
43245X1=1
```

```
43246 IF X<0 THEN X1=-1
43247 X3=ABS(X)
43248 Z1=(X3-1)/(X3+1)
43249 IF X3<.5 THEN Z1=X3
43250 Z=Z1*Z1
43252 GOSUB 43320
43254 GOSUB 43200
43255 Y=Z1*Y
43256 IF X3>=.5 THEN Y=Y+.78539816339745
43257 Y=X1*Y
43258 RETURN
43260 X1=1
43261 C=-1
43262 X2=10
43263 X3=X
43264 IF X>=1 THEN GOTO 43268
43265 X=1/X
43266 X1=-1
43268 X=X2*X
43269 C=C+1
43270 X=X/X2
43271 IF X>X2 THEN GOTO 43269
43272 Z1=(X-3.162277660)/(X+3.162277660)
43273 Z=Z1*Z1
43274 GOSUB 43340
43275 GOSUB 43200
43276 Y=X1*(C+Z1*Y+.5)
43277 X=X3
43278 RETURN
43280 X1=1
43281 C=-1
43282 X2=2.718281828459045
43283 X3=X
43284 IF X>=1 THEN GOTO 43288
43285 X=1/X
43286 X1=-1
43288 X=X2*X
43289 C=C+1
43290 X=X/X2
43291 IF X>X2 THEN GOTO 43289
43292 Z1=(X-1.6487212707)/(X+1.6487212707)
43293 Z=Z1*Z1
43294 GOSUB 43360
43295 GOSUB 43200
43296 Y=X1*(C+Z1*Y+.5)
43297 X=X3
43298 RETURN
43300 N=6
43301 A(0)=1
```

```
43302A(1)=-.1666666666671334
43303A(2)=.00833333333809067
43304A(3)=-.000198412715551283
43305A(4)=.0000275575897507620
43306A(5)=-.00000002507059876207
43307A(6)=.0000000000164105986683
43308RETURN
43310N=6
43311A(0)=1
43312A(1)=-.4999999999982
43313A(2)=.04166666664651
43314A(3)=-.001388888805755
43315A(4)=.000024801428034
43316A(5)=-.0000002754213324
43317A(6)=.0000000020189405
43318RETURN
43320N=10
43321A(0)=1
43322A(1)=-.333333311286
43323A(2)=.199998774421
43324A(3)=-.142831560376
43325A(4)=.110840091104
43326A(5)=-.089229124381
43327A(6)=.070315200033
43328A(7)=-.049278908030
43329A(8)=.026879941561
43330A(9)=-.009568384520
43331A(10)=.001605444922
43332RETURN
43340N=9
43341A(0)=.8685889644
43342A(1)=.2895297117
43343A(2)=.1737120608
43344A(3)=.1242584742
43345A(4)=.0939080460
43346A(5)=.1009301264
43347A(6)=-.0439630355
43348A(7)=.3920576195
43349A(8)=-.5170494708
43350A(9)=.4915571108
43351RETURN
43360N=9
43361A(0)=2
43362A(1)=.66666672443
43363A(2)=.3999895288
43364A(3)=.286436047
43365A(4)=.197959107
43366A(5)=.6283533
43367A(6)=-4.54692
```

```
43368A(7)=28.117
43369A(8)=-86.42
43370A(9)=106.1
43371RETURN
43380N=8
43381A(0)=1
43382A(1)=.99999999668
43383A(2)=.49999995173
43384A(3)=.16666704243
43385A(4)=.04166685027
43386A(5)=.00832672635
43387A(6)=.00140836136
43388A(7)=.00017358267
43389A(8)=.00003931683
43390RETURN
43400N=9
43401A(0)=1
43402A(1)=1.1512925485
43403A(2)=.6627373050
43404A(3)=.2543345675
43405A(4)=.0732032741
43406A(5)=.0168603036
43407A(6)=.0032196227
43408A(7)=.0005547662
43409A(8)=.0000573305
43410A(9)=.0000179419
43411RETURN
43450X1=1
43451X3=X
43452IFX<0THENX1=-1
43453X=ABS(X)
43455X2=INT(X)
43456X=X-X2
43458GOSUB43400
43459Z=X
43461GOSUB43200
43462Y=Y*Y
43463FORI=1TOX2
43464Y=Y*10
43465NEXTI
43466IFX1<0THENY=1/Y
43467X=X3
43468RETURN
43470X1=1
43471X3=X
43472IFX<0THENX1=-1
43473X=ABS(X)
43475X2=INT(X)
43476X=X-X2
43478GOSUB43380
43479Z=X
43481GOSUB43200
43482FORI=1TOX2
43483Y=Y*2.718281828459045
43484NEXTI
43485IFX1<0THENY=1/Y
43486X=X3
43487RETURN
```

APPENDIX III

*Conversion to Other BASICs and
Microsoft BASIC Program Listings*

APPENDIX III
Conversion to Other BASICs and Microsoft BASIC Program Listings

Presented in this appendix are aids for converting the subroutines listed in Appendix IIA to other BASICs, particularly Microsoft. As was discussed in the Preface and Chapter I, the language subset employed in the subroutines requires little revision to allow those programs to be used with most BASIC interpreters. Most of the changes are apparent from a consideration of the alterations needed for compatibility with the very popular Microsoft dialect. The following items are specific to North Star-to-Microsoft translation. Only the subroutines (line numbers 40000 and up) are treated.

Also included in this appendix is a program listing of a nearly universal Microsoft version (CP/M-based). Although there are differences between the Microsoft BASIC dialects as used in the Altair, Apple, PET, TRS-80, and other computers, they are all close enough in their essentials so that this listing should be applicable with few changes.

The North Star "," and Microsoft ";"

In North Star BASIC, the use of a comma in a print statement permits free format listing with no carriage returns until the end of the line is reached. "Free format" means that the items listed are printed immediately next to one another on the line, with a separating space in the case of numeric fields. The use of a comma in Microsoft BASIC leads to a tabular listing. Each numerical field, for example, starts in a particular column. The Microsoft equivalent of the North Star comma is the semicolon.

The Random Number Generator, RND

The difference between the North Star and Microsoft RND functions is important and is discussed in Chapter V. In North Star BASIC, using RND(X) with $0 < X < 1$ sets a new seed, and with $X = 0$ the random number sequence in progress is simply continued. With Microsoft BASIC, in contrast, the use of $X < 0$ usually sets a new seed, and the sequence in progress is continued with $X > 0$. $X = 0$ generally results in the return of the last number generated. To translate between the two BASICs, the RND(0) North Star BASIC should be changed to RND(0.999). The particular value chosen for the argument, 0.999, works with all known (to the author) versions of Microsoft BASIC. The RND function appears in lines:

42927	42950	42977	43000	43025
43050	43076	43105	43150	

FOR/NEXT Loops

The North Star BASIC interpreter compares the loop argument with the limit *before* the loop is executed. The Microsoft interpreter performs the comparison *after* the loop is executed. Thus, all FOR/NEXT loops are processed *at least once* in Microsoft BASIC, even if the argument is beyond the limit. For example, the loop starting with "FOR I = 7 TO 5" will be skipped in North Star BASIC and executed once in Microsoft. To avoid this seemingly illogical behavior in Microsoft BASIC, test the loop variable with an "IF" statement before the "FOR" statement. This test has been included in several places in the Microsoft listing given in this appendix. For example, see lines 40208 to 40211.

The Square Root Function

The square root function in North Star BASIC is SQRT( ), whereas in Microsoft BASIC it is SQR( ). The translation is trivial, but important. See lines 40400, 41704, 41761, 42706, and 42722.

The TAB Function

North Star and Microsoft BASIC are on the whole consistent in their use of the TAB function. However, the Applesoft dialect contains an idiosyncrasy; TAB(0) causes a carriage shift to column 255. Thus, TAB(0) should be avoided. The TAB function appears in lines:

40016	40032	40038	40047	40049	40123
40139	40145	40156	40158	40166	40169
40172					

General Information

North Star and Microsoft BASIC contain several important functions which are used in the subroutines in this book, but which may not be available in other interpreters. However, most of these functions are supplied in Chapter VI as subroutines or can be derived from those programs.

Sine and Cosine

The SINE and COSINE subroutines given in Chapter VI may be used to replace the functions SIN( ) and COS( ). These functions are used in the following statements:

SIN( ):	40451	41400	41401	43051
COS( ):	40450	41400	41402	43051

Arctangent

The arctangent (inverse tangent) function is not included in many BASICs, including North Star Version 6, Release 3 or earlier versions. Subroutine ARCTAN can be used to replace the ATN( ) function which appears in the following statements: 40407 and 41762.

Other Functions

Other functions used by the subroutines which are available in both North Star and Microsoft BASIC, but which may not be present in smaller interpreters, are the natural logarithm (LOG( )), the exponent (EXP( )), and the power (^). The first two may be replaced by the LN(X) and EXP(X) subroutines given in Chapter VI. The third function can be generated by observing that $Y^X = EXP(X*LOG(Y))$. The statement numbers at which these functions appear are shown below:

LOG( ):	43001	43028	43029	43077	43151
EXP( ):	42950	43027	43029		
^ :	40402	40404	41100	41150	41760
	43107	43100	43151		

CP/M Microsoft BASIC Program Listings

The following is a concatenated listing of the CP/M Microsoft BASIC versions of the demonstration programs and subroutines presented in the text. This particular dialect is very closely compatible with all Microsoft extended BASIC interpreters, including the TRS-80 (Level II), Apple II (Applesoft), PET/CBM, and Altair. Small incompatibilities are the double precision indicators "+" and "D". The former is automatically added by the CP/M version to the end of fixed-point format double-precision numbers such as 1.234567891011#. Similarly, "D" replaces "E" in double-precision exponential notation numbers such as 0.1234567891011D-10. If your BASIC does not support this notation, remove "#" and change "D" to "E".

The size of the concatenated Microsoft BASIC demonstration programs is 12727 bytes. The associated subroutine library size is 21500 bytes. Note that both of these are significantly larger than the corresponding North Star BASIC values given in Appendix IIA.

```
100 PRINT"PROGRAM FOR PLOTTING EQUALLY"
101 PRINT"SPACED DATA VALUES"
102 PRINT
103 PRINT"THE USER INPUTS THE STARTING "
104 PRINT"AND ENDING COORDINATES, ALONG "
105 PRINT"WITH THE NUMBER OF EVALUATION POINTS."
106 PRINT"THE PROGRAM WILL THEN PLOT "
107 PRINT"THE DATA"
108 PRINT
109 PRINT
110 REM INITIALIZATION
111 PRINT "INPUT TERMINAL WIDTH: ";
112 INPUT L
113 PRINT "BEGINNING CORDINATE: ";
114 INPUT X1
115 PRINT "ENDING COORDINATE: ";
116 INPUT X2
117 PRINT "NUMBER OF DATA POINTS: ";
118 INPUT N
119 DIM D(N+1)
120 REM DATA INPUT
121 PRINT"INPUT DATA"
122 FOR I=1 TO N
123 PRINT I;
124 INPUT D(I)
125 NEXT I
126 REM GO TO PLOTTING SUBROUTINE PROPER
127 GOSUB 40000
128 END
129 PRINT"PROGRAM FOR PLOTTING EQUALLY"
130 PRINT"SPACED FUNCTION VALUES"
131 PRINT
132 PRINT"THE USER INPUTS THE STARTING "
133 PRINT"AND ENDING COORDINATES, ALONG "
134 PRINT"WITH THE NUMBER OF EVALUATION POINTS."
135 PRINT"THE PROGRAM WILL THEN PLOT "
136 PRINT"THE FUNCTION OVER THAT RANGE."
137 PRINT
138 PRINT
139 REM INITIALIZATION
140 PRINT "INPUT TERMINAL WIDTH: ";
141 INPUT L
142 PRINT "BEGINNING CORDINATE: ";
143 INPUT X1
144 PRINT "ENDING COORDINATE: ";
145 INPUT X2
146 PRINT "NUMBER OF DATA POINTS: ";
147 INPUT N
148 DIM D(N+1)
```

```
149 REM FUNCTION EVALUATION
150 FOR I=1 TO N
151 REM *****INPUT FUNCTION BELOW*****
152 X=X1+(I-1)*(X2-X1)/(N-1)
153 D(I)=.1*X*X*X-3*X*X+2*X-3
154 NEXT I
155 REM GO TO PLOTTING SUBROUTINE PROPER
156 GOSUB 40000
157 END
158 PRINT"PROGRAM FOR PLOTTING COORDINATE"
159 PRINT"SETS OF DATA"
160 PRINT
161 PRINT"THE PROGRAM ASSUMES THAT THE DATA IS"
162 PRINT"SEQUENTIALLY ORDERED, FIRST BY ABSCISSA"
163 PRINT"VALUE, SECOND BY ORDINATE"
164 PRINT"THE PROGRAM WILL THEN PLOT "
165 PRINT"THE DATA."
166 PRINT
167 PRINT
168 REM INITIALIZATION
169 PRINT "INPUT TERMINAL WIDTH: ";
170 INPUT L
171 PRINT "NUMBER OF DATA POINTS: ";
172 INPUT N
173 DIM D(N+1),C(N+1),E(N+1)
174 REM DATA INPUT
175 PRINT "INPUT DATA IN ABSCISSA, ORDINATE PAIRS: "
176 FOR I=1 TO N
177 PRINT I;
178 INPUT C(I),D(I)
179 REM C(I)=ABSCISSA, D(I)=ORDINATE
180 NEXT I
181 REM GO TO PLOTTING SUBROUTINE PROPER
182 GOSUB 40100
183 END
184 REM PROGRAM TO DEMONSTRATE COMPLEX NUMBER
185 REM ADDITION, SUBTRACTION, MULTIPLICATION
186 REM AND DIVISION
187 PRINT"ENTER COMPLEX NUMBERS IN (X,Y) PAIRS:"
188 PRINT
189 PRINT"Z1= ";
190 INPUT X(1),Y(1)
191 PRINT
192 PRINT"Z2= ";
193 INPUT X(2),Y(2)
194 REM COMPLEX NUMBER ADDITION
195 GOSUB 40300
196 PRINT
197 PRINT
```

```
198 PRINT "Z1+Z2= ";X(3);
199 IF Y(3)>=0 THEN PRINT"+";
200 PRINT Y(3);" I"
201 PRINT
202 PRINT
203 REM COMPLEX NUMBER SUBTRACTION
204 GOSUB 40350
205 PRINT "Z1-Z2= ";X(3);
206 IF Y(3)>=0 THEN PRINT"+";
207 PRINT Y(3);" I"
208 PRINT
209 REM COMPLEX NUMBER MULTIPLICATION
210 GOSUB 40600
211 PRINT"Z1*Z2= ";X;
212 IF Y>=0 THEN PRINT" +";
213 PRINT Y;" I"
214 PRINT
215 REM COMPLEX NUMBER DIVISION
216 GOSUB 40800
217 PRINT"Z1/Z2= ";X;
218 IF Y>=0 THEN PRINT" +";
219 PRINT Y;" I"
220 END
221 REM PROGRAM TO DEMONSTRATE
222 REM RECTANGULAR TO POLAR CONVERSION
223 PRINT"INPUT RECTANGULAR COORDINATES TO BE CONVERTED";
224 INPUT X,Y
225 REM CONVERSION
226 GOSUB 40400
227 PRINT
228 PRINT"POLAR COORDINATES ARE : RADIUS= ";U
229 PRINT"                        ANGLE=  ";V;" RADIANS"
230 END
231 REM PROGRAM TO DEMONSTRATE
232 REM POLAR TO RECTANGULAR CONVERSION
233 PRINT"INPUT POLAR COORDINATES TO BE CONVERTED";
234 INPUT U,V
235 REM CONVERSION
236 GOSUB 40450
237 PRINT
238 PRINT"RECTANGULAR COORDINATES ARE (X,Y)= ";X;" ,";Y
239 END
240 REM PROGRAM TO DEMONSTRATE
241 REM COMPLEX NUMBER TO A POWER
242 PRINT
243 PRINT"INPUT THE POWER DESIRED, FOLLOWED BY THE"
244 PRINT"COMPLEX NUMBER"
245 PRINT"N ";
246 INPUT N
```

```
247 PRINT"X ";
248 INPUT X
249 PRINT"Y ";
250 INPUT Y
251 REM POWER CALCULATION
252 GOSUB 41200
253 PRINT
254 PRINT"RESULT= ";X;
255 IF Y1>=0 THEN PRINT" +";
256 PRINT Y;" I"
257 END
258 REM PROGRAM TO DEMONSTRATE
259 REM FINDING THE ROOTS OF A COMPLEX NUMBER
260 PRINT
261 PRINT"INPUT THE INTEGER ROOT DESIRED FOLLOWED"
262 PRINT"BY THE COMPLEX NUMBER"
263 PRINT"N ";
264 INPUT N
265 PRINT"X ";
266 REM X9 AND Y9 ARE STORED VALUES
267 INPUT X9
268 PRINT"Y ";
269 INPUT Y9
270 REM FIND N ROOTS
271 PRINT
272 PRINT
273 PRINT"ORDER     X+ YI"
274 PRINT"-----    ----------------"
275 PRINT
276 FOR M=1 TO N
277 REM CONVERT TO X,Y
278 X=X9
279 Y=Y9
280 GOSUB 41300
281 PRINT"   ";M;
282 PRINT"   ";INT(1000*X+.5)/1000;"   ";
283 IF Y>=0 THEN PRINT "+";
284 PRINT INT(1000*Y+.5)/1000;" I"
285 NEXT M
286 END
287 REM PROGRAM TO DEMONSTRATE SPHERICAL TO
288 REM CARTESIAN COORDINATE CONVERSION
289 PRINT"INPUT RADIUS, U: ";
290 INPUT U
291 PRINT"INPUT ANGLE IN (X,Y) PLANE, V: ";
292 INPUT V
293 PRINT"INPUT ANGLE OFF Z AXIS, W: ";
294 INPUT W
295 GOSUB 41400
```

```
296 PRINT
297 PRINT"(X,Y,Z)= (";X;" , ";Y;" , ";Z;")"
298 END
299 REM PROGRAM TO DEMONSTRATE CARTESIAN TO
300 REM SPHERICAL COORDINATE CONVERSION
301 PRINT"INPUT X, Y AND Z: ";
302 INPUT X,Y,Z
303 GOSUB 41450
304 PRINT
305 PRINT"RADIUS= ";U
306 PRINT"ANGLE IN (X,Y) PLANE= ";V;" RADIANS"
307 PRINT"ANGLE OFF Z AXIS= ";W;" RADIANS"
308 END
309 REM PROGRAM TO DEMONSTRATE VECTOR SUBROUTINES
310 N=3
311 DIM A(N),B(N),C(N)
312 PRINT "INPUT VECTOR A (X,Y,Z): ";
313 INPUT A(1),A(2),A(3)
314 PRINT "INPUT VECTOR B (X,Y,Z): ";
315 INPUT B(1),B(2),B(3)
316 REM SUM
317 GOSUB 41500
318 PRINT
319 PRINT "A+B= (";C(1);", ";C(2);", ";C(3);")"
320 REM DIFFERENCE
321 GOSUB 41550
322 PRINT
323 PRINT"A-B= (";C(1);", ";C(2);", ";C(3);")"
324 REM CROSS PRODUCT
325 GOSUB 41650
326 PRINT
327 PRINT"AXB= (";C(1);", ";C(2);", ";C(3);")"
328 REM DOT PRODUCT
329 GOSUB 41600
330 PRINT
331 PRINT "A.B= ";C
332 REM VECTOR ANGLE
333 GOSUB 41750
334 PRINT
335 PRINT "ANGLE BETWEEN VECTORS A AND B= ";A;" RADIANS"
336 REM VECTOR LENGTH
337 GOSUB 41700
338 PRINT
339 PRINT "LENGTH OF VECTOR B= ";L
340 PRINT
341 END
342 REM PROGRAM TO DEMONSTRATE MATRIX OPERATIONS
343 PRINT"INPUT THE ROW AND COLUMN DIMENSIONS OF A(I,J) AND B(I,J):"
344 PRINT"ROW SIZE= ";
```

```
345 INPUT M
346 PRINT"COLUMN SIZE= ";
347 INPUT N
348 REM FIND MAXIMUM MATRIX DIMENSION NEEDED
349 IF M>N THEN K=M+1
350 IF N>=M THEN K=N+1
351 DIM A(K,K), B(K,K), C(K,K)
352 REM INPUT MATRIX A
353 PRINT
354 PRINT"INPUT MATRIX A ROW BY ROW"
355 PRINT
356 FOR I=1 TO M
357 PRINT"INPUT THE ";N;" ELEMENTS OF ROW ";I;" :"
358 FOR J=1 TO N
359 INPUT A(I,J)
360 NEXT J
361 PRINT
362 NEXT I
363 PRINT
364 REM PRINT MATRIX A
365 REM MOVE A TO C
366 N1=M
367 N2=N
368 N3=0
369 GOSUB 42150
370 PRINT"MATRIX A="
371 PRINT
372 REM PRINTING SUBROUTINE
373 GOSUB 1020
374 PRINT
375 REM INPUT MATRIX B
376 PRINT
377 PRINT"INPUT MATRIX B ROW BY ROW"
378 PRINT
379 FOR I=1 TO M
380 PRINT"INPUT THE ";N;" ELEMENTS OF ROW ";I;" :"
381 FOR J=1 TO N
382 INPUT B(I,J)
383 NEXT J
384 PRINT
385 NEXT I
386 PRINT
387 REM MOVE B TO C
388 GOSUB 42125
389 PRINT
390 PRINT"MATRIX B="
391 PRINT
392 REM PRINTING SUBROUTINE
393 GOSUB 1020
```

```
394 PRINT
395 REM MATRIX ADDITION
396 GOSUB 41800
397 PRINT"A+B="
398 REM PRINTING SUBROUTINE
399 GOSUB 1020
400 REM MATRIX SUBTRACTION
401 GOSUB 41850
402 PRINT"A-B="
403 REM PRINTING SUBROUTINE
404 GOSUB 1020
405 REM MATRIX TRANSPOSE (A)
406 REM FIRST, SAVE B IN C
407 GOSUB 42125
408 REM PUT TRANSPOSE OF A IN B
409 GOSUB 41950
410 REM MOVE TRANSPOSE OF A TO A
411 N1=N
412 N2=M
413 GOSUB 42075
414 REM RETURN C TO B
415 N1=M
416 N2=N
417 GOSUB 42100
418 PRINT
419 PRINT"A(TRANSPOSE)="
420 REM SWITCH ROW AND COLUMN SIZES FOR THE TRANSPOSE PRINT
421 N2=M
422 N1=N
423 M=N1
424 N=N2
425 REM MOVE A TO C FOR PRINTING
426 GOSUB 42150
427 PRINT
428 REM PRINTING SUBROUTINE
429 GOSUB 1020
430 REM DETERMINE PRODUCT OF A(TRANSPOSE) AND B
431 M1=M
432 N1=N
433 M2=N
434 N2=M
435 REM MATRIX MULTIPLICATION SUBROUTINE
436 GOSUB 41900
437 PRINT"A(TRANSPOSE) X B="
438 PRINT
439 REM PRINTING SUBROUTINE
440 N=M
441 GOSUB 1020
442 PRINT
```

```
443 END
444 REM PROGRAM TO DEMONSTRATE SPECIAL MATRIX OPERATIONS
445 REM FIRST CREATE A DIAGONAL MATRIX
446 PRINT"WHAT SIZE MATRIX IS DESIRED";
447 INPUT N
448 PRINT
449 PRINT"THIS MATRIX WILL HAVE ONLY DIAGONAL ELEMENTS.":
450 PRINT" WHAT VALUE SHOULD THEY BE";
451 INPUT B
452 DIM A(N+1,N+1),B(N+1,N+1),C(N+1,N+1)
453 REM GO TO DIAGONAL MATRIX CREATION SUBROUTINE
454 GOSUB 42000
455 REM SAVE MATRIX IN BOTH A AND C
456 N1=N
457 N2=N
458 N3=0
459 GOSUB 42075
460 GOSUB 42125
461 REM PRINT RESULT
462 M=N
463 GOSUB 1020
464 REM SWITCH TWO ROWS
465 PRINT
466 PRINT"INPUT THE NUMBERS OF THE TWO ROWS TO BE SWITCHED";
467 INPUT N1,N2
468 REM GO TO SWITCHING SUBROUTINE
469 GOSUB 42250
470 REM MOVE RESULT TO C AND PRINT
471 N1=N
472 N2=N
473 GOSUB 42150
474 PRINT
475 GOSUB 1020
476 PRINT"INPUT SCALE FACTOR TO BE MULTIPLIED BY: ";
477 INPUT B
478 GOSUB 42200
479 REM PRINT RESULT
480 GOSUB 42150
481 GOSUB 1020
482 PRINT
483 PRINT"INPUT WHAT FRACTION OF ROW N1 IS TO BE ADDED TO ROW N2"
484 PRINT"FRACTION: ";
485 INPUT B
486 PRINT"N1: ";
487 INPUT N1
488 PRINT"N2: ";
489 INPUT N2
490 REM GO TO OPERATION
491 GOSUB 42275
```

```
492 REM PRINT RESULT
493 PRINT
494 N2=N
495 N1=N
496 GOSUB 42150
497 GOSUB 1020
498 PRINT"CLEAR MATRIX"
499 N1=N
500 N2=N
501 GOSUB 42225
502 REM PRINT RESULT
503 PRINT
504 GOSUB 42150
505 GOSUB 1020
506 END
507 REM PROGRAM TO DEMONSTRATE DETERMINANT
508 PRINT"INPUT MATRIX SIZE: ";
509 INPUT N
510 DIM A(N+1,N+1),B(N+1,N+1),C(N+1,N+1)
511 PRINT
512 REM INPUT MATRIX
513 FOR I=1 TO N
514 PRINT"INPUT ROW ";I
515 FOR J=1 TO N
516 INPUT A(I,J)
517 NEXT J
518 PRINT
519 NEXT I
520 PRINT
521 PRINT "MATRIX A="
522 REM MOVE A TO C FOR PRINTING
523 N1=N
524 N2=N
525 N3=0
526 GOSUB 42150
527 REM PRINT
528 M=N
529 GOSUB 1020
530 PRINT
531 REM EVALUATE DETERMINANT
532 GOSUB 42350
533 PRINT"DETERMINANT= ";D
534 END
535 REM PROGRAM TO DEMONSTRATE MATRIX INVERSION
536 PRINT"INPUT DIMENSION OF MATRIX TO BE INVERTED: ";
537 INPUT N
538 DIM A(N+1,N+1),B(N+1,2*N+2),C(N+1,N+1)
539 PRINT
540 FOR I=1 TO N
```

```
541 PRINT"INPUT ROW ";I;" :"
542 FOR J=1 TO N
543 INPUT A(I,J)
544 NEXT J
545 PRINT
546 NEXT I
547 PRINT
548 PRINT"MATRIX A="
549 N1=N
550 N2=N
551 N3=0
552 REM MOVE A TO C
553 GOSUB 42150
554 M=N
555 GOSUB 1020
556 REM GOTO INVERSION SUBROUTINE
557 GOSUB 42400
558 REM MOVE B TO C TO PRINT
559 PRINT"INVERSE OF A="
560 PRINT
561 GOSUB 42125
562 M=N
563 GOSUB 1020
564 REM CHECK RESULTS
565 REM MULTIPLY A TIMES B
566 M1=N
567 M2=N
568 GOSUB 41900
569 M=N
570 REM PRINT RESULTS
571 PRINT
572 PRINT"MATRIX A TIMES INVERSE MATRIX A="
573 GOSUB 1020
574 END
575 REM PROGRAM TO DEMONSTRATE USE OF INVERSE TO SOLVE
576 REM SIMULTANEOUS EQUATIONS
577 PRINT"INPUT SIZE OF EQUATION (NUMBER OF UNKNOWNS): ";
578 INPUT N
579 DIM A(N+1,N+1),B(N+1,2*N+2),C(N+1,N+1)
580 PRINT
581 FOR I=1 TO N
582 PRINT"INPUT ROW ";I;" OF COEFFICIENT MATRIX"
583 FOR J=1 TO N
584 INPUT A(I,J)
585 NEXT J
586 PRINT
588 PRINT"INPUT CONSTANT WHICH ROW EQUATION EQUALS: ";
589 INPUT B(I,1)
590 PRINT
```

```
591 NEXT I
592 PRINT
593 PRINT"MATRIX A="
594 N1=N
595 N2=N
596 N3=0
597 REM MOVE A TO C
598 GOSUB 42150
599 REM PRINT A
600 M=N
601 GOSUB 1020
603 REM MOVE CONSTANT VECTOR TO C
604 GOSUB 42125
605 REM PRINT CONSTANT VECTOR
606 PRINT
607 PRINT"CONSTANT VECTOR ="
608 N=1
609 GOSUB 1020
610 N=M
611 REM OBTAIN INVERSE
612 GOSUB 42400
613 REM MOVE RESULT IN B TO A
614 GOSUB 42075
615 REM MOVE CONSTANT VECTOR IN C TO B
616 GOSUB 42100
617 REM MULTIPLY INVERSE TIMES CONSTANT VECTOR
618 M1=N
619 N1=N
620 M2=N
621 N2=1
622 GOSUB 41900
623 REM RESULT IS IN C. PRINT C
624 M=N
625 N=1
626 PRINT"SOLUTION VECTOR="
627 GOSUB 1020
628 END
629 REM PROGRAM TO DEMONSTRATE OBTAINING THE LARGEST EIGENVALUE OF A MATRIX
630 PRINT"WHAT IS THE SIZE OF THE MATRIX: ";
631 INPUT N
632 PRINT
633 DIM A(N+1,N+1),B(N+1,N+1),C(N+1,N+1)
634 FOR I=1 TO N
635 PRINT"INPUT ROW ";I
636 FOR J=1 TO N
637 INPUT A(I,J)
638 NEXT J
639 PRINT
```

```
640 NEXT I
641 PRINT
642 PRINT"INPUT LEVEL OF ACCURACY: ";
643 INPUT E
644 PRINT"INPUT NUMBER OF ITERATIONS AT WHICH TO STOP: ";
645 INPUT D1
646 REM FIND EIGENVALUE
647 GOSUB 42700
648 PRINT
649 PRINT"EIGENVALUE= ";A
650 PRINT
651 PRINT"EIGENVECTOR="
652 FOR I=1 TO N
653 PRINT B(I,1)
654 NEXT I
655 PRINT
656 PRINT "NUMBER OF ITERATIONS= ";D
657 END
658 REM PROGRAM TO DEMONSTRATE THE MATRIX EXPONENT SUBROUTINE
659 PRINT"INPUT THE SIZE OF THE MATRIX TO BE EXPONENTIATED";
660 INPUT N
661 DIM A(N+1,N+1), B(N+1,N+1), C(N+1,N+1), D(N+1,N+1)
662 FOR I=1 TO N
663 PRINT"INPUT ROW ";I
664 FOR J=1 TO N
665 INPUT A(I,J)
666 NEXT J
667 PRINT
668 NEXT I
669 PRINT
670 REM PRINT MATRIX A
671 N1=N
672 N2=N
673 N3=0
674 M=N
675 PRINT"MATRIX A="
676 PRINT
677 REM MOVE A TO C
678 GOSUB 42150
679 REM PRINT
680 GOSUB 1020
681 PRINT"INPUT THE VARIABLE X";
682 INPUT X
683 PRINT
684 PRINT"INPUT THE NUMBER OF TERMS TO BE CALCULATED";
685 INPUT K2
686 REM OBTAIN EXPONENT
687 GOSUB 42800
688 PRINT
```

```
689 PRINT"EXP(AX)= "
690 PRINT
691 GOSUB 1020
692 PRINT
693 END
694 REM PROGRAM TO DEMONSTRATE THE USE OF THE UNIFORM
695 REM RANDOM NUMBER GENERATOR
696 L=80
697 REM SET PLOT WIDTH
698 N=21
699 DIM C(22),D(22),E(22)
700 FOR I=1 TO 21
701 D(I)=0
702 C(I)=(I-1)/10
703 NEXT I
704 REM SET SEED
705 Z2=RND(-.5)
706 FOR I2=1 TO 1000
707 Z2=10*RND(.999)+1
708 Z2=INT(Z2)
709 D(Z2)=D(Z2)+1
710 NEXT I2
711 PRINT"DISTRIBUTION OF THE 1000 RANDOM NUMBERS"
712 PRINT"IN TERMS OF THE NUMBER PER INTERVAL OF LENGTH 1/10"
713 PRINT
714 PRINT
715 GOSUB 40100
716 PRINT
717 PRINT
718 END
719 REM SINE AND COSINE APPROXIMATION DEMONSTRATION
720 PRINT"   X         SIN(X)             DELTA              COS(X)
    DELTA"
721 PRINT"  ---        --------          -------            -------
    -------"
722 PRINT
723 FOR X=-5 TO 5 STEP .2
724 GOSUB 43210
725 PRINT INT(10*X)/10;TAB(8);Y;TAB(26);(SIN(X)-Y);
726 GOSUB 43225
727 PRINT TAB(46);Y;TAB(65);(COS(X)-Y)
728 NEXT X
729 END
730 REM ARCTANGENT APPROXIMATION DEMONSTRATION
731 PRINT" X";TAB(8);"ARCTANGENT";TAB(24);"DELTA"
732 PRINT"---"TAB(8);"----------";TAB(24);"-----"
733 PRINT
734 FOR X=-5 TO 5 STEP .2
735 GOSUB 43245
```

```
736 PRINT X;TAB(8);Y;TAB(21);(ATN(X)-Y)
737 NEXT X
738 END
739 REM NATURAL LOGARITHM AND EXPONENT APPROXIMATION DEMONSTRATION
740 PRINT" X";TAB(8);"   LN(X)   ";TAB(24);" DELTA ";TAB(40);"EXP(LN(X))";
TAB(56);" DELTA "
741 PRINT"---";TAB(8);"---------":TAB(24);"-------";TAB(40);"----------"
;TAB(56);"-------"
742 FOR X=.1 TO 5 STEP .1
743 GOSUB 43280
744 PRINT X;TAB(8);Y;TAB(24);(LOG(X)-Y);
745 F=X
746 X=Y
747 GOSUB 43470
748 PRINT TAB(40);Y;TAB(56);(F-Y)
749 X=F
750 NEXT X
751 END
752 REM BASE 10 LOGARITHM AND EXPONENT APPROXIMATION DEMONSTRATION
753 PRINT" X ";TAB(8);" LOG(X) ";TAB(24);" DELTA ";TAB(40);"10^(LOG(X))"
;TAB(56);" DELTA "
754 PRINT"---";TAB(8);"--------";TAB(24);"-------";TAB(40);"-----------"
;TAB(56);"-------"
755 FOR X=.1 TO 5 STEP .1
756 GOSUB 43260
757 PRINT X;TAB(8);Y;TAB(24);(.4342944#*LOG(X)-Y);
758 F=X
759 X=Y
760 GOSUB 43450
761 PRINT TAB(40);Y;TAB(56);(F-Y)
762 X=F
763 NEXT X
764 END
1000 REM *******************
1010 REM PRINTING SUBROUTINE
1020 PRINT
1030 FOR I=1 TO M
1040 FOR J=1 TO N
1050 PRINT TAB(6*J);INT(C(I,J)*100+.5)/100;
1060 NEXT J
1070 PRINT
1080 NEXT I
1090 PRINT
1100 RETURN
```

APPENDIX III

Microsoft Listing of Subroutines

```
39998 REM PLOTTING SUBROUTINE (EQAPLOT)
39999 REM SHIFT DATA TO NON-NEGATIVE
40000 B=100000000#
40001 REM FIND MINIMUM DATA VALUE
40002 FOR I=1 TO N
40003 IF B>D(I) THEN B=D(I)
40004 NEXT I
40005 REM SUBTRACT MINIMUM VALUE FROM ALL DATA
40006 FOR I=1 TO N
40007 D(I)=D(I)-B
40008 NEXT I
40009 REM FIND MAX. SHIFTED DATA VALUE
40010 C=0
40011 FOR I=1 TO N
40012 IF C<D(I) THEN C=D(I)
40013 NEXT I
40014 REM DETERMINE PRINTING SCALE VALUE
40015 A=L/C
40016 REM FIND TAB POSITION OF ZERO
40017 E=A*ABS(B)
40018 PRINT
40019 PRINT
40020 PRINT"***** DATA PLOT (SCALED) *****"
40021 PRINT
40022 PRINT
40023 PRINT"MIN. ORDINATE=   ";B;"    MAX. ORDINATE= ";C+B
40024 PRINT"INITIAL ABSCISSA VALUE= ";X1
40025 PRINT
40026 PRINT
40027 REM IF B IS POSITIVE, SKIP ZERO LABEL
40028 IF B>0 THEN GOTO 40034
40029 REM IF DATA ARE ALL BELOW ZERO, SKIP LABEL
40030 IF ABS(B)>C THEN GOTO 40034
40031 REM LABEL ZERO
40032 PRINT TAB(E);"0"
40033 REM GO TO AXIS PRINT SUBROUTINE
40034 GOSUB 40200
40035 FOR I=1 TO N
40036 REM INSERT LINE FEED FOR AUTO SPACING
40037 IF INT(.6*L/N)<1 THEN GOTO 40041
40038 FOR K=1 TO INT(.6*L/N):PRINT ":";TAB(L);":"
40039 NEXT K
40040 REM LOCATE DATUM POSITION
40041 E2=A*D(I)
40042 REM FORMATTED PRINT
40043 IF E2>=1 THEN GOTO 40046
40044 PRINT"*";
40045 GOTO 40049
40046 PRINT":";
```

```
40047 PRINT TAB(E2);"*";
40048 IF INT(E2)=L THEN GOTO 40050
40049 PRINT TAB(L);":";
40050 PRINT
40051 NEXT I
40052 REM GO TO AXIS PRINT SUBROUTINE
40053 GOSUB 40200
40054 PRINT
40055 PRINT
40056 PRINT"END ABSCISSA VALUE= ";X2
40057 PRINT
40058 PRINT
40059 REM RETURN TO DATA SOURCE PROGRAM
40060 RETURN
40098 REM TWO DIMENSIONAL DATA PLOTTING SUBROUTINE (DATAPLOT)
40099 REM SHIFT DATA TO NON-NEGATIVE
40100 B=100000000#
40101 REM FIND MINIMUM DATA VALUE
40102 FOR I=1 TO N
40103 IF B>D(I) THEN B=D(I)
40104 NEXT I
40105 REM SUBTRACT MINIMUM VALUE FROM ALL DATA
40106 FOR I=1 TO N
40107 D(I)=D(I)-B
40108 NEXT I
40109 REM FIND MAX. SHIFTED DATA VALUE
40110 C=0
40111 FOR I=1 TO N
40112 IF C<D(I) THEN C=D(I)
40113 NEXT I
40114 REM DETERMINE E(I), THE ABSCISSA SPACINGS
40115 E(0)=0
40116 E(N)=1
40117 FOR I=2 TO N
40118 E(I-1)=INT(.5*(C(I)-C(I-1))*L/(C(N)-C(1))+.5)
40119 REM SPACING SCALED ACCORDING TO LINE WIDTH
40120 NEXT I
40121 REM DETERMINE PRINTING SCALE VALUE
40122 A=L/C
40123 REM FIND TAB POSITION OF ZERO
40124 E=A*ABS(B)
40125 PRINT
40126 PRINT
40127 PRINT"***** DATA PLOT (SCALED) *****"
40128 PRINT
40129 PRINT
40130 PRINT"MIN. ORDINATE=   ";B;"    MAX. ORDINATE= ";C+B
40131 PRINT"INITIAL ABSCISSA VALUE= ";C(1)
40132 PRINT
```

```
40133 PRINT
40134 REM IF B IS POSITIVE, SKIP ZERO LABEL
40135 IF B>0 THEN GOTO 40141
40136 REM IF DATA ARE ALL BELOW ZERO, SKIP LABEL
40137 IF ABS(B)>C THEN GOTO 40141
40138 REM LABEL ZERO
40139 PRINT TAB(E);"0"
40140 REM GO TO AXIS PRINT SUBROUTINE
40141 GOSUB 40200
40142 FOR I=1 TO N
40143 REM INSERT FEED FOR ABSCISSA SPACING
40144 IF E(I-1)<1 THEN GOTO 40148
40145 FOR K=1 TO E(I-1):PRINT":";TAB(L);":"
40146 NEXT K
40147 REM LOCATE DATUM POSITION
40148 E2=A*D(I)
40149 REM TEST FOR MULTIPLE ORDINATE ABSCISSA
40150 IF E(I)=0 THEN GOTO 40161
40151 REM FORMATTED PRINT
40152 IF E2>=1 THEN GOTO 40155
40153 PRINT"*";
40154 GOTO 40158
40155 PRINT":";
40156 PRINT TAB(E2);"*";
40157 IF INT(E2)=L THEN GOTO 40159
40158 PRINT TAB(L);":";
40159 PRINT
40160 GOTO 40175
40161 REM TEST FOR OVERLAY OF POINTS
40162 REM IF DUPLICATE POINT, SKIP TO END
40163 IF D(I)=D(I+1) THEN GOTO 40175
40164 IF E2>=1 THEN GOTO 40168
40165 REM PRINT DOUBLE VALUES ON ONE LINE
40166 PRINT"*";TAB(A*D(I+1));"*";
40167 GOTO 40171
40168 PRINT":";
40169 PRINT TAB(E2);"*";TAB(A*D(I+1));"*";
40170 REM TEST IF LINE LIMIT REACHED
40171 IF INT(A*D(I+1))=L THEN GOTO 40173
40172 PRINT TAB(L);":";
40173 PRINT
40174 I=I+1
40175 NEXT I
40176 REM GO TO AXIS PRINT SUBROUTINE
40177 GOSUB 40200
40178 PRINT
40179 PRINT
40180 PRINT"END ABSCISSA VALUE= ";C(N)
40181 PRINT
```

```
40182 PRINT
40183 REM RETURN TO DATA SOURCE PROGRAM
40184 RETURN
40199 REM AXIS PLOT (AXISPLOT)
40200 E3=E-5*INT(E/5)
40201 REM IF B IS POSTIVE, THEN SKIP ZERO LABEL
40202 IF B>0 THEN E3=0
40203 REM IF B IS GREATER THAN THE LARGEST VALUE, SKIP
40204 IF ABS(B)>C THEN E3=0
40205 IF E3<1 THEN GOTO 40208
40206 FOR J=1 TO E3:PRINT"-";
40207 NEXT J
40208 IF (L-E3)/5<1 THEN GOTO 40211
40209 FOR J=1 TO (L-E3)/5:PRINT"I----";
40210 NEXT J
40211 PRINT"I";
40212 E4=(J-1)*5+1+E3
40213 IF E4=L+1 THEN PRINT
40214 IF E4=L+1 THEN GOTO 40221
40215 E4=E4+1
40216 IF E4>=L+1 THEN GOTO 40219
40217 PRINT"-";
40218 GOTO 40215
40219 PRINT":"
40220 REM RETURN TO MAIN PLOTTING PROGRAM
40221  RETURN
40299 REM COMPLEX NUMBER ADDITION SUBROUTINE (ZADD)
40300 X(3)=X(1)+X(2)
40301 Y(3)=Y(1)+Y(2)
40302 RETURN
40349 REM COMPLEX NUMBER SUBTRACTION SUBROUTINE (ZSUB)
40350 X(3)=X(1)-X(2)
40351 Y(3)=Y(1)-Y(2)
40352 RETURN
40399 REM RECTANGULAR TO POLAR CONVERSION SUBROUTINE (RECT/POL)
40400 U=SQR(X*X+Y*Y)
40401 REM GUARD AGAINST AMBIGUOUS VECTOR
40402 IF Y=0 THEN Y=(.1)^30
40403 REM GUARD AGAINST DIVIDE BY ZERO
40404 IF X=0 THEN X=(.1)^30
40405 REM SOME BASICS REQUIRE A SIMPLE ARGUMENT
40406 W=Y/X
40407 V=ATN(W)
40408 REM CHECK QUADRANT AND ADJUST
40409 IF X<0 THEN V=V+3.1415926535#
40410 IF V<0 THEN V=V+6.2831853072#
40411 RETURN
40449 REM POLAR TO RECTANGULAR CONVERSION SUBROUTINE (POL/RECT)
40450 X=U*COS(V)
```

```
40451 Y=U*SIN(V)
40452 RETURN
40499 REM POLAR MULTIPLICATION SUBROUTINE (ZPOLMLT)
40500 U=U(1)*U(2)
40501 V=V(1)+V(2)
40502 IF V>=6.2831853072# THEN V=V-6.2831853072#
40503 RETURN
40549 REM POLAR DIVISION SUBROUTINE (ZPOLDIV)
40550 U=U(1)/U(2)
40551 V=V(1)-V(2)
40552 IF V<0 THEN V=V+6.2831853072#
40553 RETURN
40599 REM RECTANGULAR COMPLEX NUMBER MULTIPLICATION SUBROUTINE (ZRECTMLT)
40600 X=X(1)
40601 Y=Y(1)
40602 REM RECTANGULAR TO POLAR CONVERSION
40603 GOSUB 40400
40604 U(1)=U
40605 V(1)=V
40606 X=X(2)
40607 Y=Y(2)
40608 REM RECTANGULAR TO POLAR CONVERSION
40609 GOSUB 40400
40610 U(2)=U
40611 V(2)=V
40612 REM POLAR MULTIPLICATION
40613 GOSUB 40500
40614 REM POLAR TO RECTANGULAR CONVERSION
40615 GOSUB 40450
40616 RETURN
40799 REM RECTANGULAR COMPLEX NUMBER DIVISION SUBROUTINE (ZRECTDIV)
40800 X=X(1)
40801 Y=Y(1)
40802 REM RECTANGULAR TO POLAR CONVERSION
40803 GOSUB 40400
40804 U(1)=U
40805 V(1)=V
40806 X=X(2)
40807 Y=Y(2)
40808 REM RECTANGULAR TO POLAR CONVERSION
40809 GOSUB 40400
40810 U(2)=U
40811 V(2)=V
40812 REM POLAR COMPLEX NUMBER DIVISION
40813 GOSUB 40550
40814 REM POLAR TO RECTANGULAR CONVERSION
40815 GOSUB 40450
40816 RETURN
41099 REM POLAR POWER SUBROUTINE (ZPOLPOW)
```

```
41100 U1=U^N
41101 V1=N*V
41102 V1=V1-6.2831853072#*INT(V1/6.2831853072#)
41103 RETURN
41149 REM POLAR (FIRST) ROOT SUBROUTINE (ZPOLRT)
41150 U1=U^(1/N)
41151 V1=V/N
41152 RETURN
41198 REM RECTANGULAR COMPLEX NUMBER POWER SUBROUTINE (ZRECTPOW)
41199 REM RECTANGULAR TO POLAR CONVERSION
41200 GOSUB 40400
41201 REM POLAR POWER
41202 GOSUB 41100
41203 REM CHANGE VARIABLE FOR CONVERSION
41204 U=U1
41205 V=V1
41206 REM POLAR TO RECTANGULAR CONVERSION
41207 GOSUB 40450
41208 RETURN
41298 REM RECTANGULAR COMPLEX NUMBER ROOT SUBROUTINE (ZRECTRT)
41299 REM RECTANGULAR TO POLAR CONVERSION
41300 GOSUB 40400
41301 REM POLAR (FIRST) ROOT
41302 GOSUB 41150
41303 U=U1
41304 REM FIND M ORDER ROOT
41305 REM M=1 CORRESPONDS TO THE FIRST ROOT
41306 V=V1+6.2831853072#*(M-1)/N
41307 REM POLAR TO RECTANGULAR CONVERSION
41308 GOSUB 40450
41309 RETURN
41399 REM SPHERICAL TO RECTANGULAR (CARTESIAN) CONVERSION SUBROUTINE (SPR/RECT)
41400 X=U*(SIN(W))*COS(V)
41401 Y=U*(SIN(W))*SIN(V)
41402 Z=U*COS(W)
41403 RETURN
41448 REM RECTANGULAR (CARTESIAN) TO SPHERICAL CONVERSION SUBROUTINE (RECT/SPR)
41449 REM RECTANGULAR TO POLAR CONVERSION
41450 GOSUB 40400
41451 REM SAVE AND CHANGE VARIABLES
41452 V1=V
41453 X=U
41454 Y=Z
41455 REM RECTANGULAR TO POLAR CONVERSION
41456 GOSUB 40400
41457 IF V>1.5707963268# THEN V=V-6.28318553072#
41458 W=1.5707963268#-V
41459 V=V1
41460 RETURN
```

```
41498 REM VECTOR ADDITION SUBROUTINE (VECTADD)
41499 REM C=A+B
41500 FOR I=1 TO N
41501 C(I)=A(I)+B(I)
41502 NEXT I
41503 RETURN
41548 REM VECTOR SUBTRACTION SUBROUTINE (VECTSUB)
41549 REM C=A-B
41550 FOR I=1 TO N
41551 C(I)=A(I)-B(I)
41552 NEXT I
41553 RETURN
41598 REM VECTOR DOT PRODUCT SUBROUTINE (VECTDOT)
41599 REM C=A.B
41600 C=0
41601 FOR I=1 TO N
41602 C=C+A(I)*B(I)
41603 NEXT I
41604 RETURN
41648 REM VECTOR CROSS PRODUCT SUBROUTINE (VECTCURL)
41649 REM C=A X B
41650 C(1)=A(2)*B(3)-A(3)*B(2)
41651 C(2)=A(3)*B(1)-A(1)*B(3)
41652 C(3)=A(1)*B(2)-A(2)*B(1)
41653 RETURN
41699 REM VECTOR LENGTH SUBROUTINE (VECTLEN)
41700 L=0
41701 FOR I=1 TO N
41702 L=L+A(I)*A(I)
41703 NEXT I
41704 L=SQR(L)
41705 RETURN
41747 REM VECTOR ANGLE SUBROUTINE (VECTANGL)
41748 REM ANGLE BETWEEN A AND B
41749 REM FIND DOT PRODUCT
41750 GOSUB 41600
41751 REM FIND LENGTH OF A
41752 GOSUB 41700
41753 REM SAVE VALUE
41754 L1=L
41755 REM FIND LENGTH OF B
41756 FOR I=1 TO N
41757 A(I)=B(I)
41758 NEXT I
41759 GOSUB 41700
41760 E=C/(L*L1)+(.1)^30
41761 E=SQR(1-E*E)/E
41762 A=ATN(E)
41763 IF C<0 THEN A=3.1415926536#-A
```

```
41764 RETURN
41798 REM MATRIX ADDITION SUBROUTINE (MATADD)
41799 REM C=A+B
41800 FOR I=1 TO M
41801 FOR J=1 TO N
41802 C(I,J)=A(I,J)+B(I,J)
41803 NEXT J
41804 NEXT I
41805 RETURN
41848 REM MATRIX SUBTRACTION SUBROUTINE (MATSUB)
41849 REM C=A-B
41850 FOR I=1 TO M
41851 FOR J=1 TO N
41852 C(I,J)=A(I,J)-B(I,J)
41853 NEXT J
41854 NEXT I
41855 RETURN
41898 REM MATRIX MULTIPLICATION SUBROUTINE (MATMULT)
41899 REM C=A X B    A IS M1 BY N1    B IS M2 BY N2    C IS M1 BY N2
41900 FOR I=1 TO M1
41901 FOR J=1 TO N2
41902 C(I,J)=0
41903 FOR K=1 TO N1
41904 C(I,J)=C(I,J)+A(I,K)*B(K,J)
41905 NEXT K
41906 NEXT J
41907 NEXT I
41908 RETURN
41948 REM MATRIX TRANSPOSE SUBROUTINE (MATTRANS)
41949 REM B=TRANSPOSE(A)
41950 FOR I=1 TO N
41951 FOR J=1 TO M
41952 B(I,J)=A(J,I)
41953 NEXT J
41954 NEXT I
41955 RETURN
41998 REM DIAGONAL MATRIX CREATION SUBROUTINE (MATDIAG)
41999 REM MATRIX B(I,J) IS THE IDENTITY MATRIX TIMES B
42000 FOR I=1 TO N
42001 FOR J=1 TO N
42002 B(I,J)=0
42003 IF I=J THEN B(I,J)=B
42004 NEXT J
42005 NEXT I
42006 RETURN
42048 REM MATRIX SAVE (A IN B) SUBROUTINE (MATSAVAB)
42049 REM N1,N2 AND N3 ARE INPUT INDICES
42050 IF N1*N2*N3=0 THEN GOTO 42060
42051 REM CHECK DIMENSION
```

```
42052 FOR I1=1 TO N1
42053 FOR I2=1 TO N2
42054 FOR I3=1 TO N3
42055 B(I1,I2,I3)=A(I1,I2,I3)
42056 NEXT I3
42057 NEXT I2
42058 NEXT I1
42059 RETURN
42060 IF N1*N2=0 THEN GOTO 42067
42061 FOR I1=1 TO N1
42062 FOR I2=1 TO N2
42063 B(I1,I2)=A(I1,I2)
42064 NEXT I2
42065 NEXT I1
42066 RETURN
42067 IF N1=0 THEN RETURN
42068 FOR I1=1 TO N1
42069 B(I1)=A(I1)
42070 NEXT I1
42071 RETURN
42073 REM MATRIX SAVE (B IN A) SUBROUTINE (MATSAVBA)
42074 REM N1,N2 AND N3 ARE INPUT INDICES
42075 IF N1*N2*N3=0 THEN GOTO 42085
42076 REM CHECK DIMENSION
42077 FOR I1=1 TO N1
42078 FOR I2=1 TO N2
42079 FOR I3=1 TO N3
42080 A(I1,I2,I3)=B(I1,I2,I3)
42081 NEXT I3
42082 NEXT I2
42083 NEXT I1
42084 RETURN
42085 IF N1*N2=0 THEN GOTO 42092
42086 FOR I1=1 TO N1
42087 FOR I2=1 TO N2
42088 A(I1,I2)=B(I1,I2)
42089 NEXT I2
42090 NEXT I1
42091 RETURN
42092 IF N1=0 THEN RETURN
42093 FOR I1=1 TO N1
42094 A(I1)=B(I1)
42095 NEXT I1
42096 RETURN
42098 REM MATRIX SAVE (C IN B) SUBROUTINE (MATSAVCB)
42099 REM N1,N2 AND N3 ARE INPUT INDICES
42100 IF N1*N2*N3=0 THEN GOTO 42110
42101 REM CHECK DIMENSION
42102 FOR I1=1 TO N1
```

```
42103 FOR I2=1 TO N2
42104 FOR I3=1 TO N3
42105 B(I1,I2,I3)=C(I1,I2,I3)
42106 NEXT I3
42107 NEXT I2
42108 NEXT I1
42109 RETURN
42110 IF N1*N2=0 THEN GOTO 42117
42111 FOR I1=1 TO N1
42112 FOR I2=1 TO N2
42113 B(I1,I2)=C(I1,I2)
42114 NEXT I2
42115 NEXT I1
42116 RETURN
42117 IF N1=0 THEN RETURN
42118 FOR I1=1 TO N1
42119 B(I1)=C(I1)
42120 NEXT I1
42121 RETURN
42123 REM MATRIX SAVE (B IN C) SUBROUTINE (MATSAVBC)
42124 REM N1,N2 AND N3 ARE INPUT INDICES
42125 IF N1*N2*N3=0 THEN GOTO 42135
42126 REM CHECK DIMENSION
42127 FOR I1=1 TO N1
42128 FOR I2=1 TO N2
42129 FOR I3=1 TO N3
42130 C(I1,I2,I3)=B(I1,I2,I3)
42131 NEXT I3
42132 NEXT I2
42133 NEXT I1
42134 RETURN
42135 IF N1*N2=0 THEN GOTO 42142
42136 FOR I1=1 TO N1
42137 FOR I2=1 TO N2
42138 C(I1,I2)=B(I1,I2)
42139 NEXT I2
42140 NEXT I1
42141 RETURN
42142 IF N1=0 THEN RETURN
42143 FOR I1=1 TO N1
42144 C(I1)=B(I1)
42145 NEXT I1
42146 RETURN
42148 REM MATRIX SAVE (A IN C) SUBROUTINE (MATSAVAC)
42149 REM N1,N2 AND N3 ARE INPUT INDICES
42150 IF N1*N2*N3=0 THEN GOTO 42160
42151 REM CHECK DIMENSION
42152 FOR I1=1 TO N1
42153 FOR I2=1 TO N2
```

```
42154 FOR I3=1 TO N3
42155 C(I1,I2,I3)=A(I1,I2,I3)
42156 NEXT I3
42157 NEXT I2
42158 NEXT I1
42159 RETURN
42160 IF N1*N2=0 THEN GOTO 42167
42161 FOR I1=1 TO N1
42162 FOR I2=1 TO N2
42163 C(I1,I2)=A(I1,I2)
42164 NEXT I2
42165 NEXT I1
42166 RETURN
42167 IF N1=0 THEN RETURN
42168 FOR I1=1 TO N1
42169 C(I1)=A(I1)
42170 NEXT I1
42171 RETURN
42173 REM MATRIX SAVE (C IN A) SUBROUTINE (MATSAVCA)
42174 REM N1,N2 AND N3 ARE INPUT INDICES
42175 IF N1*N2*N3=0 THEN GOTO 42185
42176 REM CHECK DIMENSION
42177 IF N1*N2=0 THEN GOTO 42192
42178 FOR I2=1 TO N2
42179 FOR I3=1 TO N3
42180 A(I1,I2,I3)=C(I1,I2,I3)
42181 NEXT I3
42182 NEXT I2
42183 NEXT I1
42184 RETURN
42186 FOR I1=1 TO N1
42187 FOR I2=1 TO N2
42188 A(I1,I2)=C(I1,I2)
42189 NEXT I2
42190 NEXT I1
42191 RETURN
42192 IF N1=0 THEN RETURN
42193 FOR I1=1 TO N1
42194 A(I1)=C(I1)
42195 NEXT I1
42196 RETURN
42198 REM SCALAR B X MATRIX A SUBROUTINE (MATSCALE)
42199 REM N1,N2 AND N3 ARE INPUT INDICES
42200 IF N1*N2*N3=0 THEN GOTO 42210
42201 REM CHECK DIMENSION
42202 FOR I1=1 TO N1
42203 FOR I2=1 TO N2
42204 FOR I3=1 TO N3
42205 A(I1,I2,I3)=B*A(I1,I2,I3)
```

```
42206 NEXT I3
42207 NEXT I2
42208 NEXT I1
42209 RETURN
42210 IF N1*N2=0 THEN GOTO 42217
42211 FOR I1=1 TO N1
42212 FOR I2=1 TO N2
42213 A(I1,I2)=B*A(I1,I2)
42214 NEXT I2
42215 NEXT I1
42216 RETURN
42217 IF N1=0 THEN RETURN
42218 FOR I1=1 TO N1
42219 A(I1)=B*A(I1)
42220 NEXT I1
42221 RETURN
42223 REM MATRIX A CLEAR SUBROUTINE (MATCLRA)
42224 REM N1,N2 AND N3 ARE INPUT INDICES
42225 IF N1*N2*N3=0 THEN GOTO 42235
42226 REM CHECK DIMENSION
42227 FOR I1=1 TO N1
42228 FOR I2=1 TO N2
42229 FOR I3=1 TO N3
42230 A(I1,I2,I3)=0
42231 NEXT I3
42232 NEXT I2
42233 NEXT I1
42234 RETURN
42235 IF N1*N2=0 THEN GOTO 42242
42236 FOR I1=1 TO N1
42237 FOR I2=1 TO N2
42238 A(I1,I2)=0
42239 NEXT I2
42240 NEXT I1
42241 RETURN
42242 IF N1=0 THEN RETURN
42243 FOR I1=1 TO N1
42244 A(I1)=0
42245 NEXT I1
42246 RETURN
42248 REM ROW SWITCHING SUBROUTINE (MATSWCH)
42249 REM ROWS N1 AND N2 ARE INTERCHANGED
42250 FOR J=1 TO N
42251 B=A(N1,J)
42252 A(N1,J)=A(N2,J)
42253 A(N2,J)=B
42254 NEXT J
42255 RETURN
42273 REM ROW MULTIPLICATION/ADD SUBROUTINE (MATRMAD)
```

```
42274 REM B TIMES ROW N1 ADDED TO N2
42275 FOR J=1 TO N
42276 A(N2,J)=A(N2,J)+B*A(N1,J)
42277 NEXT J
42278 RETURN
42296 REM COFACTOR K SUBROUTINE (MATCOFAT)
42297 REM INPUT MATRIX SIZE IS N X N
42298 REM MATRIX A(I,J) IN, MATRIX B(I,J) OUT
42299 REM FIRST SHIFT UP ONE ROW
42300 FOR I=2 TO N
42301 FOR J=1 TO N
42302 B(I-1,J)=A(I,J)
42303 NEXT J
42304 NEXT I
42305 FOR I=1 TO N-1
42306 FOR J=K TO N
42307 IF K=N THEN GOTO 42309
42308 B(I,J)=B(I,J+1)
42309 NEXT J
42310 NEXT I
42311 RETURN
42348 REM MATRIX DETERMINANT SUBROUTINE (MATDET)
42349 REM FINDS DETERMINANT FOR UP TO A 4 X 4 MATRIX
42350 IF N>=2 THEN GOTO 42355
42351 REM ********************
42352 REM FIRST ORDER DETERMINANT
42353 D=A(1,1)
42354 RETURN
42355 IF N>=3 THEN GOTO 42360
42356 REM ********************
42357 REM SECOND ORDER DETERMINANT
42358 D=A(1,1)*A(2,2)-A(1,2)*A(2,1)
42359 RETURN
42360 IF N>=4 THEN GOTO 42370
42361 REM ********************
42362 REM THIRD ORDER DETERMINANT
42363 D=A(1,1)*(A(2,2)*A(3,3)-A(2,3)*A(3,2))
42364 D=D-A(1,2)*(A(2,1)*A(3,3)-A(2,3)*A(3,1))
42365 D=D+A(1,3)*(A(2,1)*A(3,2)-A(2,2)*A(3,1))
42366 RETURN
42367 REM ********************
42368 REM FOURTH ORDER DETERMINANT
42369 REM SAVE A IN C
42370 N1=N
42371 N2=N
42372 N3=0
42373 GOSUB 42150
42374 IF N>=5 THEN RETURN
42375 REM D1 WILL BE THE DETERMINANT
```

```
42376 D1=0
42377 REM FIND DETERMINANT OF EACH COFACTOR
42378 FOR K=1 TO 4
42379 REM GET COFACTOR K
42380 GOSUB 42300
42381 REM COFACTOR RETURNED IN B
42382 REM MOVE B TO A
42383 GOSUB 42075
42384 REM GET DET(A)
42385 GOSUB 42363
42386 D1=D1+C(1,K)*D
42387 REM REVERSE SIGN FOR NEXT COFACTOR
42388 D1=-D1
42389 REM SAVE C IN A
42390 GOSUB 42175
42391 NEXT K
42392 D=D1
42393 RETURN
42395 REM MATRIX INVERSION SUBROUTINE (MATINV)
42396 REM GAUSS-JORDAN ELIMINATION
42397 REM MATRIX A IS INPUT, MATRIX B IS OUTPUT
42398 REM DIM A=N X N     TEMPORARY DIM B=N X 2N
42399 REM FIRST CREATE MATRIX WITH A ON THE LEFT AND I ON THE RIGHT
42400 FOR I=1 TO N
42401 FOR J=1 TO N
42402 B(I,J+N)=0
42403 B(I,J)=A(I,J)
42404 NEXT J
42405 B(I,I+N)=1
42406 NEXT I
42407 REM PERFORM ROW ORIENTED OPERATIONS TO CONVERT THE LEFT HAND
42408 REM SIDE OF B TO THE IDENTITY MATRIX. THE INVERSE OF A WILL
42409 REM THEN BE ON THE RIGHT.
42410 FOR K=1 TO N
42411 IF K=N THEN GOTO 42424
42412 M=K
42413 REM FIND MAXIMUM ELEMENT
42414 FOR I=K+1 TO N
42415 IF ABS(B(I,K))>ABS(B(M,K)) THEN M=I
42416 NEXT I
42417 IF M=K THEN GOTO 42424
42418 FOR J=K TO 2*N
42419 B=B(K,J)
42420 B(K,J)=B(M,J)
42421 B(M,J)=B
42422 NEXT J
42423 REM DIVIDE ROW K
42424 FOR J=K+1 TO 2*N
42425 B(K,J)=B(K,J)/B(K,K)
```

```
42426 NEXT J
42427 IF K=1 THEN GOTO 42434
42428 FOR I=1 TO K-1
42429 FOR J=K+1 TO 2*N
42430 B(I,J)=B(I,J)-B(I,K)*B(K,J)
42431 NEXT J
42432 NEXT I
42433 IF K=N THEN GOTO 42441
42434 FOR I=K+1 TO N
42435 FOR J=K+1 TO 2*N
42436 B(I,J)=B(I,J)-B(I,K)*B(K,J)
42437 NEXT J
42438 NEXT I
42439 NEXT K
42440 REM RETRIEVE INVERSE FROM THE RIGHT SIDE OF B
42441 FOR I=1 TO N
42442 FOR J=1 TO N
42443 B(I,J)=B(I,J+N)
42444 NEXT J
42445 NEXT I
42446 RETURN
42693 REM EIGENVALUE (POWER METHOD) SUBROUTINE (EIGENPOW)
42694 REM AX=LX
42695 REM A IS THE N X N MATRIX
42696 REM B IS AN ARBITRARY VECTOR
42697 REM E IS THE RELATIVE ERROR CHOSEN
42698 REM  D= COUNT OF THE NUMBER OF ITERATIONS
42699 REM SET PARAMETERS NEEDED FOR MULTIPLY SUBROUTINE
42700 M1=N
42701 N1=N
42702 M2=N
42703 N2=1
42704 REM GENERATE ARBITRARY NORMALIZED VECTOR B(I,1)
42705 FOR I=1 TO N
42706 B(I,1)=1/SQR(N)
42707 NEXT I
42708 REM B = LAST EIGENVALUE ESTIMATE
42709 REM A = CURRENT EIGENVALUE ESTIMATE
42710 REM PICK AN INITIAL VALUE FOR THE EIGENVALUE GUESS
42711 B=1
42712 D=0
42713 REM START ITERATION
42714 A=0
42715 GOSUB 41900
42716 REM CONVERT C OUTPUT TO B
42717 FOR I=1 TO N
42718 B(I,1)=C(I,1)
42719 A=A+B(I,1)*B(I,1)
42720 NEXT I
```

```
42721 D=D+1
42722 A=SQR(A)
42723 REM NORMALIZE VECTOR
42724 FOR I=1 TO N
42725 B(I,1)=B(I,1)/A
42726 NEXT I
42727 IF ABS((A-B)/A)<E THEN RETURN
42728 B=A
42729 IF D>D1 THEN RETURN
42730 GOTO 42714
42796 REM MATRIX EXPONENT SUBROUTINE (MATEXP)
42797 REM INPUTS TO THE SUBROUTINE ARE THE MATRIX A, MATRIX
42798 REM SIZE N, NUMBER OF TERMS K2, AND VARIABLE X
42799 REM SET UP INDICES TO BE USED LATER
42800 N1=N
42801 N2=N
42802 N3=0
42803 M1=N
42804 M2=N
42805 REM GUARD AGAINST DIVIDE BY ZERO
42806 IF X=0 THEN X=1E-13
42807 REM INITIALIZE STORAGE MATRIX D(I,J)
42808 FOR I=1 TO N
42809 FOR J=1 TO N
42810 D(I,J)=0
42811 NEXT J
42812 NEXT I
42813 REM K2 IS THE NUMBER OF TERMS TO BE CALCULATED
42814 K1=0
42815 REM CREATE IDENTITY MATRIX IN B
42816 B=1
42817 GOSUB 42000
42818 REM MOVE B TO C
42819 GOSUB 42125
42820 REM ADD TO D
42821 GOSUB 42847
42822 K1=K1+1
42823 IF K1>=K2 THEN GOTO 42838
42824 REM SCALE MATRIX A BY X/K1
42825 B=X/K1
42826 GOSUB 42200
42827 REM MULTIPLY A TIMES B
42828 GOSUB 41900
42829 REM ADD RESULT TO MATRIX D
42830 GOSUB 42847
42831 REM MOVE C TO B
42832 GOSUB 42100
42833 REM RETURN MATRIX A TO ORIGINAL CONDITION
42834 B=K1/X
```

```
42835 GOSUB 42200
42836 REM CONTINUE SUMMATION
42837 GOTO 42822
42838 REM MOVE RESULT IN D TO C
42839 FOR I=1 TO N
42840 FOR J=1 TO N
42841 C(I,J)=D(I,J)
42842 NEXT J
42843 NEXT I
42844 REM RETURN TO CALLING PROGRAM
42845 RETURN
42846 REM D(I,J) IS USED FOR TEMPORARY STORAGE
42847 FOR I=1 TO N
42848 FOR J=1 TO N
42849 D(I,J)=D(I,J)+C(I,J)
42850 NEXT J
42851 NEXT I
42852 RETURN
42899 REM LINEAR RANDOM NUMBER GENERATOR (LINEAR)
42900 REM U=MEAN, V=SPREAD, D=SEED
42901 I9=I9+1
42902 A=3.1415926535889793#
42903 B=2.718281828459045#
42904 C=1.414213562373095#
42905 D=1+ABS(D)
42906 E=E+(1+D/B)*C
42907 E=E*I9
42908 E=E-A*INT(E/A)
42909 E=E-INT(E)+.018
42910 IF E>.1 THEN E=E+9E-03
42911 IF E>.2 THEN E=E-2E-03
42912 IF E>.3 THEN E=E-5E-03
42913 IF E>.4 THEN E=E-5E-03
42914 IF E>.5 THEN E=E-.015
42915 E=V*(E-.5)+U
42916 RETURN
42923 REM NORMAL DISTRIBUTION BY CENTRAL LIMIT THEOREM (NORMAL)
42924 REM U=MEAN, V=STANDARD DEVIATION, E=RANDOM NO. GENERATED
42925 E=0
42926 FOR I9=1 TO 48
42927 E=E+RND(.999)-.5
42928 NEXT I9
42929 E=V*E/2+U
42930 RETURN
42948 REM POISSON RANDOM NUMBER GENERATOR (POISSON)
42949 REM INPUT PARAMETER U
42950 X=RND(.999)*EXP(U)
42951 X1=1
42952 Y1=1
```

```
42953 Y=0
42954 IF X1>X THEN GOTO 42959
42955 Y=Y+1
42956 Y1=Y1*U/Y
42957 X1=X1+Y1
42958 GOTO 42954
42959 IF Y>0 THEN Y=Y-(X1-X)/Y1
42960 E=Y
42961 RETURN
42972 REM BINOMIAL RANDOM NUMBER GENERATOR (BINOMIAL)
42973 REM B=PROBABILITY PER TRIAL
42974 REM N=NUMBER OF TRIALS
42975 E=0
42976 FOR K=1 TO N
42977 Y1=RND(.999)
42978 IF Y1<B THEN E=E+1
42979 NEXT K
42980 RETURN
42998 REM EXPONENTIAL RANDOM NUMBER GENERATOR (EXPONENT)
42999 REM U=MEAN
43000 X=RND(.999)
43001 E=-U*LOG(1-X)
43002 RETURN
43023 REM FERMI RANDOM NUMBER GENERATOR (FERMI)
43024 REM U=INFLECTION POINT, V=SPREAD OF TRANSITION REGION
43025 X=RND(.999)
43026 Y=1
43027 A=EXP(4*U/V)
43028 B=(X-1)*LOG(1+A)
43029 Y1=B+LOG(A+EXP(Y))
43030 IF ABS((Y-Y1)/Y)<1E-03 THEN GOTO 43033
43031 Y=Y1
43032 GOTO 43029
43033 E=V*Y1/4
43034 RETURN
43048 REM CAUCHY RANDOM NUMBER GENERATOR (CAUCHY)
43049 REM U=MEAN
43050 X=RND(.999)
43051 E=U*SIN(1.5707963267#*X)/COS(1.5707963267#*X)
43052 RETURN
43073 REM GAMMA (N=2) RANDOM NUMBER GENERATOR (GAMMA)
43074 REM B=INPUT PARAMETER
43075 Y=1
43076 X=RND(.999)
43077 Y1=-LOG((1-X)/(1+Y))
43078 IF ABS((Y1-Y)/Y)<1E-03 THEN GOTO 43081
43079 Y=Y1
43080 GOTO 43077
43081 E=B*Y1
```

```
43082 RETURN
43096 REM BETA RANDOM NUMBER GENERATOR (BETA)
43097 REM INPUT PARAMETERS ARE A AND B
43098 REM A IS RESTRICTED TO A=1 AND A=2
43099 REM GUARD AGAINST DIVIDE BY ZERO
43100 IF B>0 THEN GOTO 43103
43101 E=1
43102 RETURN
43103 REM B>0
43104 IF A>2 THEN RETURN
43105 X=RND(.999)
43106 IF A=2 THEN GOTO 43109
43107 E=1-(1-X)^(1/B)
43108 RETURN
43109 Y=1
43110 Y1=1-((1-X)/(1+B*Y))^(1/B)
43111 IF ABS((Y-Y1)/Y)<1E-03 THEN GOTO 43114
43112 Y=Y1
43113 GOTO 43110
43114 E=Y1
43115 RETURN
43148 REM WEIBULL RANDOM NUMBER GENERATOR (WEIBULL)
43149 REM INPUT PARAMETERS ARE U AND V
43150 X=RND(.999)
43151 E=U*((LOG(1/(1-X)))^(1/V))
43152 RETURN
43199 REM SERIES SUMMATION SUBROUTINE (SERSUM)
43200 Y=0
43201 FOR I=N TO 0 STEP -1
43202 Y=Z*Y+A(I)
43203 NEXT I
43204 RETURN
43209 REM SINE SERIES SUBROUTINE (SINE)
43210 X1=1
43211 IF X<0 THEN X1=-1
43212 X3=ABS(X)
43213 X2=3.141592653589793#
43214 REM REDUCE RANGE
43215 X3=X3-2*X2*INT(.5*X3/X2)
43216 IF X3>X2 THEN X1=-X1
43217 IF X3>X2 THEN X3=X3-X2
43218 IF X3>X2/2 THEN X3=X2-X3
43219 Z=X3*X3
43220 GOSUB 43300
43221 GOSUB 43200
43222 Y=X1*X3*Y
43223 RETURN
43224 REM COSINE SERIES SUBROUTINE (COSINE)
43225 X2=3.141592653589793#
```

```
43226 X1=1
43227 REM REDUCE RANGE
43228 X3=ABS(X)
43229 X3=X3-2*X2*INT(.5*X3/X2)
43230 IF X3>X2 THEN X1=-1
43231 IF X3>X2 THEN X3=X3-X2
43232 IF X3>X2/2 THEN X1=-X1
43233 IF X3>X2/2 THEN X3=X2-X3
43234 Z=X3*X3
43235 GOSUB 43310
43236 GOSUB 43200
43237 Y=X1*Y
43238 RETURN
43244 REM ARCTANGENT SERIES SUBROUTINE (ARCTAN)
43245 X1=1
43246 IF X<0 THEN X1=-1
43247 X3=ABS(X)
43248 Z1=(X3-1)/(X3+1)
43249 IF X3<.5 THEN Z1=X3
43250 Z=Z1*Z1
43251 REM GET SERIES COEFFICIENTS
43252 GOSUB 43320
43253 REM SUM SERIES
43254 GOSUB 43200
43255 Y=Z1*Y
43256 IF X3>=.5 THEN Y=Y+.78539816339745#
43257 Y=X1*Y
43258 RETURN
43259 REM LOG BASE TEN SERIES SUBROUTINE (LOG(X))
43260 X1=1
43261 C=-1
43262 X2=10
43263 X3=X
43264 IF X>=1 THEN GOTO 43268
43265 X=1/X
43266 X1=-1
43267 REM REDUCE RANGE
43268 X=X2*X
43269 C=C+1
43270 X=X/X2
43271 IF X>X2 THEN GOTO 43269
43272 Z1=(X-3.16227766#)/(X+3.16227766#)
43273 Z=Z1*Z1
43274 GOSUB 43340
43275 GOSUB 43200
43276 Y=X1*(C+Z1*Y+.5)
43277 X=X3
43278 RETURN
43279 REM NATURAL LOGARITHM SERIES SUBROUTINE (LN(X))
```

```
43280 X1=1
43281 C=-1
43282 X2=2.718281828459045#
43283 X3=X
43284 IF X>=1 THEN GOTO 43288
43285 X=1/X
43286 X1=-1
43287 REM REDUCE RANGE
43288 X=X2*X
43289 C=C+1
43290 X=X/X2
43291 IF X>X2 THEN GOTO 43289
43292 Z1=(X-1.6487212707#)/(X+1.6487212707#)
43293 Z=Z1*Z1
43294 GOSUB 43360
43295 GOSUB 43200
43296 Y=X1*(C+Z1*Y+.5)
43297 X=X3
43298 RETURN
43299 REM SINE SERIES COEFFICIENTS (SINDATA)
43300 N=6
43301 A(0)=1
43302 A(1)=-.1666666666671334#
43303 A(2)=8.33333333809067D-03
43304 A(3)=-1.984127155551283D-04
43305 A(4)=2.75575897507620D-06
43306 A(5)=-2.507059876207D-08
43307 A(6)=1.64105986683D-10
43308 RETURN
43309 REM COSINE SERIES COEFFICIENTS (COSDATA)
43310 N=6
43311 A(0)=1
43312 A(1)=-.4999999999982#
43313 A(2)=.04166666664651#
43314 A(3)=-1.388888805755D-03
43315 A(4)=2.4801428034D-05
43316 A(5)=-2.754213324D-07
43317 A(6)=2.0189405D-09
43318 RETURN
43319 REM ARCTANGENT SERIES COEFFICIENTS (ARCDATA)
43320 N=10
43321 A(0)=1
43322 A(1)=-.333333311286#
43323 A(2)=.199998774421#
43324 A(3)=-.142831560376#
43325 A(4)=.110840091104#
43326 A(5)=-.089229124381#
43327 A(6)=.070315200033#
43328 A(7)=-.04927890803#
```

```
43329 A(8)=.026879941561#
43330 A(9)=-9.56838452D-03
43331 A(10)=1.605444922D-03
43332 RETURN
43339 REM LOG BASE TEN SERIES COEFFICIENTS (LOGDATA)
43340 N=9
43341 A(0)=.8685889644#
43342 A(1)=.2895297117#
43343 A(2)=.1737120608#
43344 A(3)=.1242584742#
43345 A(4)=.093908046#
43346 A(5)=.1009301264#
43347 A(6)=-.0439630355#
43348 A(7)=.3920576195#
43349 A(8)=-.5170494708#
43350 A(9)=.4915571108#
43351 RETURN
43359 REM NATURAL LOGARITHM SERIES COEFFICIENTS (LNDATA)
43360 N=9
43361 A(0)=2
43362 A(1)=.66666672443#
43363 A(2)=.3999895288#
43364 A(3)=.286436047#
43365 A(4)=.197959107#
43366 A(5)=.628353
43367 A(6)=-4.54692
43368 A(7)=28.117
43369 A(8)=-86.42
43370 A(9)=106.1
43371 RETURN
43379 REM POWER OF E SERIES COEFFICIENTS (EXPDATA)
43380 N=8
43381 A(0)=1
43382 A(1)=.99999999668#
43383 A(2)=.49999995173#
43384 A(3)=.16666704243#
43385 A(4)=.04166685027#
43386 A(5)=8.32672635D-03
43387 A(6)=1.40836136D-03
43388 A(7)=1.7358267D-04
43389 A(8)=3.93168E-05
43390 RETURN
43399 REM POWER OF TEN SERIES COEFFICIENTS (TENDATA)
43400 N=9
43401 A(0)=1
43402 A(1)=1.1512925485#
43403 A(2)=.6627373050000001#
43404 A(3)=.2543345675#
43405 A(4)=.0732032741#
```

```
43406 A(5)=.0168603036#
43407 A(6)=3.2196227D-03
43408 A(7)=5.54766E-04
43409 A(8)=5.73305E-05
43410 A(9)=1.79419E-05
43411 RETURN
43449 REM POWER OF TEN SERIES SUBROUTINE (TENPOW)
43450 X1=1
43451 X3=X
43452 IF X<0 THEN X1=-1
43453 X=ABS(X)
43454 REM REDUCE RANGE
43455 X2=INT(X)
43456 X=X-X2
43457 REM GET COEFFICIENTS
43458 GOSUB 43400
43459 Z=X
43460 REM SUM SERIES
43461 GOSUB 43200
43462 Y=Y*Y
43463 IF X2<1 THEN GOTO 43466
43464 FOR I=1 TO X2:Y=Y*10
43465 NEXT I
43466 IF X1<0 THEN Y=1/Y
43467 X=X3
43468 RETURN
43469 REM EXPONENT SERIES SUBROUTINE (EXP(X))
43470 X1=1
43471 X3=X
43472 IF X<0 THEN X1=-1
43473 X=ABS(X)
43474 REM REDUCE RANGE
43475 X2=INT(X)
43476 X=X-X2
43477 REM GET COEFFICIENTS
43478 GOSUB 43380
43479 Z=X
43480 REM SUM SERIES
43481 GOSUB 43200
43482 IF X2<1 THEN GOTO 43485
43483 FOR I=1 TO X2:Y=Y*2.718281828459045#
43484 NEXT I
43485 IF X1<0 THEN Y=1/Y
43486 X=X3
43487 RETURN
```

REFERENCES

1. Daniel, C., and F. S. Wood. *Fitting Equations to Data.* Wiley-Interscience, New York, 1971.

2. Cooley, W. W., and P. R. Lohnes. *Multivariate Data Analysis.* John Wiley and Sons Inc, New York, 1971.

3. Fike, C. T. *Computer Evaluation of Mathematical Functions.* Prentice-Hall, Englewood Cliffs NJ, 1968.

4. Smith, J. M. *Scientific Analysis on the Pocket Calculator.* Wiley-Interscience, New York, 1975.

5. Gilbert, J. *Advanced Applications for Pocket Calculators.* Tab Books, Blue Ridge Summit PA, 1975.

6. Luke, Y. L. *The Special Functions and Their Approximations,* Volume I. Academic Press, New York, 1969.

7. Beckett, R., and J. Hurt. *Numerical Calculations and Algorithms.* McGraw-Hill, New York, 1967.

8. Kuo, S. *Computer Applications of Numerical Methods.* Addison-Wesley, Reading MA, 1972.

9. Smith, J.M. *Mathematical Modeling and Digital Simulation for Engineers and Scientists.* Wiley-Interscience, New York, 1977.

10. Luke, Y.L. *The Special Functions and Their Approximations,* Volume II. Academic Press, New York, 1969.

11. Sharpe, F.S., and N.L. Jacob. *BASIC.* Collier Books, New York, 1971.

12. *H.P.-25 Applications Programs.* Hewlett-Packard, 1975.

13. *H.P.-55 Statistics Programs.* Hewlett-Packard, 1974.

14. *General Statistics.* Monroe International, Orange NJ, 1964.

15. Hubin, W.N. *BASIC Programming for Scientists and Engineers.* Prentice-Hall Inc, Englewood Cliffs NJ, 1978.

16. Rogowski, S. J. *Problems for Computer Solution.* Educomp, Hartford CT, 1975.

17. Poole, L., and M. Borchers. *Some Common BASIC Programs.* Adam Osborne and Associates, Berkeley CA, 1977.

18. McCracken, D. *Fortran with Engineering Applications.* John Wiley and Sons Inc, New York, 1967.

19. Singer, B. M. *Programming in BASIC, with Applications.* McGraw-Hill, New York, 1973.

20. Schick, W., and C. J. Merz. *Fortran for Engineering.* McGraw-Hill, New York, 1972.

21. Acton, F. S. *Numerical Methods That Work.* Harper and Row, New York, 1970.

22. Carnahan, B., H. A. Luther, and J. O. Wilkes. *Applied Numerical Methods.* John Wiley and Sons Inc, New York, 1969.

23. Carnahan, B., and J. O. Wilkes. "Applied Numerical Methods" (class notes). University of Michigan, 1978.

24. Nie, N. H., C. H. Hull, J. G. Jenkins, K. Steinbrenner, and D. H. Bent. *Statistical Package for the Social Sciences.* McGraw-Hill, New York, 1975.

25. Tracton, K. *57 Practical Programs and Games in BASIC.* Tab Books, Blue Ridge Summit PA, 1978.

26. *SR-51 User's Manual.* Texas Instruments, Dallas TX, 1974.

27. Kreysig, E. *Advanced Engineering Mathematics.* John Wiley and Sons Inc, New York, 1962.

28. Dwyer, T. A., and M. Critchfield. *BASIC and the Personal Computer.* Addison-Wesley, Reading MA, 1978.

29. *System/360 Scientific Subroutine Package.* Version III, Fourth Edition. IBM Corporation, White Plains NY, 1968.

30. Abramowitz, M., and I. A. Stegun. *Handbook of Mathematical Functions.* Dover Publications Inc, New York, 1965.

31. Barnett, E. H. *Programming Time-Shared Computers in BASIC.* Wiley-Interscience, New York, 1972.

32. Crow, E. L., F. A. Davis, and M. W. Maxfield. *Statistics Manual.* Dover Publications Inc, New York, 1960.

33. Feller, W. *An Introduction to Probability Theory and Its Applications*, Volume I. John Wiley and Sons Inc, New York, 1957.

34. Ostle, B. *Statistics in Research.* Iowa State University Press, Ames IA, 1963.

35. Tracton, K. *Programs in BASIC for Electronic Engineers, Technicians and Experimenters.* Tab Books, Blue Ridge Summit PA, 1979.

36. Ruckdeschel, F. R. "Functional Approximations." *BYTE,* November 1978.

37. Hastings, C. *Approximations for Digital Computers.* Princeton University Press, Princeton NJ, 1955.

38. Ruckdeschel, F. R. "MITS vs. North Star.....which is faster?" *Kilobaud Microcomputing,* August 1978.

39. Hart, J. F., et al. *Computer Approximations.* John Wiley and Sons Inc, New York, 1968.

40. Tracton, K. *24 Tested, Ready to Run Game Programs in BASIC.* Tab Books, Blue Ridge Summit PA, 1978.

41. Ruckdeschel, F. R. "The OSI Model 500." *Kilobaud Microcomputing,* March 1979.

Index

accuracy 169
alternating series 161
amplitude 106
arctangent 168
arctan(x) 161, 165, 172
beta distribution 152
binomial distribution 134, 141
Cauchy distribution 148
Central Limit Theorem 134
characteristic polynomials 105, 108
Chebyshev polynomial expansion 167
Chi-Square distribution 134
coefficient matrix 101, 105, 111
cofactors 88
complex conjugate pairs of roots 112
complex number operations 31, 38
    addition and subtraction 31
    multiplication and division 31
complex plane 24
complex variables 24
conservation of probability 138
constitutive independent linear
    equations 101
constitutive relationships 191
convergence test 160
coordinate transformations 86
cosine 168
cos(x) 172
crystal lattice 146
cumulative distribution functions 139
data plotting 6, 16
dectiles 128
determinants 87, 107
diagonal matrix 65
diagonal matrix inversion 94
differential equations 117
differential matrix equations 118
direction cosines 57
economization 167
eigenequation 111
eigenvalues 105, 111
eigenvectors 105, 111
equations of motion 106
execution speed 2
exponent functions 168
exponential distribution 144, 150
exp(x) 167, 174
failure analysis 155

failure rate 156
Fermi distribution 146
function plotting 13
gamma distribution 150
Gauss-Jordan Elimination 94
identity matrix 66, 118
infant mortality 156
integral equations 127
iteration 109, 112, 151, 153
Kramer's rule 88
largest eigenvalue 111
least-squares regression 167
light profile 149
linear distribution 131
linearization 101
logarithm functions 168
Log-Normal distribution 135
$\log_e(x)$ 173
$\log_{10}(x)$ 166, 173
MacLaurin series 118
matrix exponentiation 117, 118
matrix inversion 94
matrix row switching 80
matrix scaling 79
matrix sums and products 63
matrix transpose 65
mean 136, 138, 141, 144, 148, 150, 153, 155
minimax 167
minimax polynomial 163
mode of oscillation 106
moments 132
Monte Carlo Analysis 125, 152
multiple eigenvalues 112
multiplicity 39
nested evaluation 169
Newton's law of motion 106
Normal distribution 134, 156
North Star RND 132
optimal expansions 166
optimal series 163
oscillations 106, 111
phase 106
Poisson distribution 137, 142
polar coordinates 25
poles 191
polynomials 108
population distribution 128
powder technology 135

power method 111
powers 38
probability density function 127
random number generators 125
range reduction 161, 169
regression analysis 108
residual 161
RND(x) 127
roots 38, 109
rotating coordinate changes 86
round-off error 161, 169
sample distribution 128
scaling coordinate changes 86
seed 127, 131, 132
shifted Gaussian distribution 134
shifting coordinate changes 86
simultaneous equations 108
simultaneous linear equations 87, 101
sine 168
singular matrix 102
sin(x) 160, 163, 165, 171
skew 150, 156
smallest eigenvalue 111
standard deviation 136, 138, 141, 144, 148, 150, 153, 155
Taylor series expansions 159
telephone switchboard simulations 144
telescoping 167
10^x 166, 174
timing comparisons 4, 171
transformation, polar 25
transpose operation 65
uniform distribution 127, 134
unweighted die 127
vectors 53
vector addition 54
vector angle 56
vector cross product 56
vector dot product 55
vector length 56
vector subtraction 54
Weibull distribution 151, 155
zeroes 191

The programs presented in this book are available on cassette and disk for most personal computers. For information on availability, contact your local computer store or:

      DYNACOMP Inc
      6 Rippingale Rd
      Pittsford NY 14534
      Phone: (716) 586-7579

Text set in Paladium Medium
by BYTE Publications

Edited by Blaise Liffick

Design and Production Supervision
by Ellen Klempner

Copy Edited by Peggy McCauley

Printed and bound using 50#MH
Matte by Halliday Lithograph Corporation,
Arcata Company, North Quincy, Massachusetts